The Flame of a Candle

The Bachelard Translation Series
Joanne H. Stroud, *Editor*
Robert S. Dupree, *Translation Editor*

WATER AND DREAMS: AN ESSAY ON THE IMAGINATION OF MATTER
LAUTRÉAMONT
AIR AND DREAMS: AN ESSAY ON THE IMAGINATION OF MOVEMENT
THE FLAME OF A CANDLE

The Flame of a Candle

GASTON BACHELARD

translated from the French by Joni Caldwell

THE BACHELARD TRANSLATIONS
THE DALLAS INSTITUTE PUBLICATIONS
THE DALLAS INSTITUTE OF HUMANITIES AND CULTURE
DALLAS

Originally published in 1961 as
La flamme d'une chandelle
Copyright 1961. Presses Universitaires, France
7th printing, copyright 1984

© 1988 by The Dallas Institute Publications

2nd Printing, copyright 2012

Cover: Drawing on paper of Gaston Bachelard by Robert Lapoujade;
collection Roger-Viollet
Design by Patricia Mora and Maribeth Lipscomb

LIBRARY OF CONGRESS CATALOGING-IN-PUBLICATION DATA

Bachelard, Gaston, 1884–1962
  (Flamme d'une chandelle. English) The Flame of a candle/by Gaston Bachelard: translated by Joni Caldwell.
    p. cm.—(The Bachelard translations)

  Translation of: La flamme d'une chandelle.
  Includes index.
  ISBN 0-911005-14-5 (alk. paper)
  ISBN 0-911005-15-3 (pbk.)(alk. paper)
    1. Flame—Meditations. I. Title. II. Series:
Bachelard, Gaston,.
1884-1962.   Bachelard translations series.
B2430.B253F5613   1988
194—dc20   89-7870
                                                   89-7870

The Dallas Institute Publications, formerly known as The Pegasus Foundation, publishes works concerned with the imaginative, mythic, and symbolic sources of culture. Publication efforts are centered at:
The Dallas Institute of Humanities and Culture
2719 Routh Street, Dallas, Texas 75201

Contents

Foreword *Joanne H. Stroud*	vii
Preface	1
1 Candles and Their Past	13
2 The Solitude of the Candle Dreamer	23
3 The Verticality of Flames	39
4 Poetic Images of the Flame in Plant Life	49
5 The Light of the Lamp	63
Epilogue	75
Endnotes	81
Author/Title Index	85
Subject Index	87

Foreword

OF THE FOUR traditional areas of poetic inspiration, Gaston Bachelard found the element of water most compatible with his psyche—more personally alluring than earth, air, or fire. Nevertheless, his probe of imagination and poetics, what initially drew his interest away from scientific inquiry, was fire imagery. Indeed, he was never satisfied that he had adequately articulated his fascination with fire. Always desiring to further plumb its depths, he returned to the subject intermittently over a thirty-year time frame. From the period of the 1930's when he first turned toward studies of the imagination and wrote *The Psychoanalysis of Fire* until his death in 1962, Bachelard was absorbed by a multiplicity of fiery images.

In *Air and Dreams* (1942) he expressed his dissatisfaction with the incompleteness of his previous work on fire. It took two more books to extend this exploration: *The Flame of a Candle*, published in 1961, and the book which his death truncated, *A Poetics of Fire*, or *The Poetics of the Phoenix*, as he sometimes called it. The latter, a collection of the fragments that he completed, has recently been published in French with a long explanatory introduction by his daughter Suzanne Bachelard. Soon to be a translation in our series, this book further cements his thematic contention that the "imagination *is* a flame, the flame of the psyche" (2, italics mine).

Bachelard's seminal philosophical and phenomenological interests are always connected to matter, to what he calls the "psychology of the familiar" (7), the rapprochement with real objects, the *friendship for things* (64). Matter sparks inner images which in turn imbue matter with memory and values. An ever renewing reciprocity of reverberations between inner and outer qualities obliterates any absolute separation between objective and subjective experience, a perspective recognized by modern physics.

Bachelard explodes the potential imprisoned in an image, in this case the candle flame. Here he emphasizes the more gentle, more radiant aspects of fire: the flame as light and soothing warmth, in contrast to the flame of lust and apocalypse which is at the opposite end of this element's spectrum. Yet, even this selected limitation gives evidence of the range of the imagination as he perceives it. The spark ignited by the energy of inspiration participates in the whole continuum of the element because fire is characteristic of the spirit, even when it sears the flesh. Bachelard's uniqueness consists of this ability to deepen awareness of the essence of fire through a multiplicity of associations and amplifications. Life cannot exist in the imagination without the image of fire as flame, Bachelard insists: "What is called *Life* in creation is, in all forms and in all beings, one and the same spirit, a single flame" (16).

It is the quality of the "livingness" of the flame that triggers Bachelard's musings of correspondences. The flame consumes and renews itself in a mysterious transformation and reformation of energies in much the way body cells do. It is the being-becoming and living-dying transposition of the flame's energy that generates clusters of images and poetic symbols for him: "The flame is thus a living substance, a poeticizing substance" (45).

Criticism was occasionally aimed at Bachelard in his lifetime for pulling solitary images out of poetic context in order to illustrate or support his point of view. But today the practice is more acceptable. Pursuing his methodology in this foreword, I have not hesitated to establish through random selection the universal, metaphorical use of fire images in precisely the manner Bachelard claims is innate to the imagination. Elemental images are the *via royale* into a "communion of imaginations" (3). His supporting examples of symbolic association draw on such poets as Novalis, Octavio Paz, and Victor Hugo. This tactic, in turn, stimulates the reader to recall other disciples of the flame whose poetic visions reverberate to the Bachelardian precept. Above all he is a teacher educating us to be more attentive readers, to be alert and sensitive to the whole fabric of poetry, claiming: "one must burn in communion with the poet" (45).

Recall, for example, William Shakespeare's "Sonnet 73" which bears witness to the flame's quintessential role as nourisher and destroyer, its life-cum-fire imagery summarized in the final quatrain:

> In me thou see'st the glowing of such fire,
> That on the ashes of his youth doth lie,
> As the deathbed whereon it must expire,
> Consumed with that which it was nourished by.

Flame as a life force also projects beyond the human into the animal world. In expanding this metaphorical association, Bachelard says that, for Novalis, "Flames constitute the very being of animal life. . . . The flame is in some way a naked animality, a kind of excessive animal. It is the glutton par excellence" (43-44). William Blake, a poet of fire, invokes flame as a metaphor for the fierceness of the life force in the untamed tiger:

> Tyger! Tyger! burning bright
> In the forests of the night,
> What immortal hand or eye
> Could frame thy fearful symmetry?
>
> In what distant deeps or skies
> Burnt the fire of thine eyes?
> On what wings dare he aspire?
> What the hand, dare seize the fire?

I know of no image that joins life, spirit, and flame at the sacred heart of all matter more compellingly than that of Gerard Manley Hopkins in his poetic celebration of the life force of a soaring bird, "The Windhover." Buttressed by Bachelard's reminder that the "flame, because it flies, is a bird" (47), we savor the magnificent revelatory image of glowing fire that climaxes the Hopkins poem:

> My heart in hiding
> Stirred for a bird,—the achieve of, the mastery of the thing!
>
> Brute beauty and valor and act, oh, air, pride, plume, here
>   Buckle! AND the fire that breaks from thee then, a billion
> Times told lovelier, more dangerous, Oh my chevalier!
>
> No wonder of it: sheér plód makes plough down sillion
> Shine, and blue-bleak embers, ah my dear,
>   Fall, gall themselves, and gash gold-vermillion.

Such fierce images of animal energy might more properly be associated with *Lautréamont*, Bachelard's exposé of instinctual cruelty. In this 1940 work, Bachelard employs the vehicle of Isidore Ducasse's poetic imagination to expose the stark savagery of fire imagery. By contrast, *The Flame of a Candle* develops images which burn slowly over an elongated span of time. He specifically connects slowly simmering fire to vegetable life. He forges a link to a vast body of poetry in which fire and arboreal nature are coupled: "poets bring them [trees and flowers] to life, to abundant poetic life through the image of flames" (10).

William Butler Yeats makes explicit Bachelard's statement that the "tree is a fire-bearer" (52). The flame is one of Yeats's most frequently used images. (I counted ninety-five instances in his poems.) To mention a random few: in "The Two Trees" the heart is imaged as a container which bears a holy tree stretching toward the infinite. The reiterated phrase, "the flaming circle of our days," together with the last verse, "the flaming circle of our life," demonstrates Yeats's distinctive use of the connective link between tree, flame, and the life force.

The tree is "the vegetable Hercules who, in every fiber of his being, prepares for his apotheosis in the flames of the pyre" (52), Bachelard cryptically reveals.

In an associated vein, the Yeatsian image of flame in its double guise, renewing—as well as destroying, burning, and self-igniting—recurs in "Two Songs from a Play," where "Whatever flames upon the night/ Man's own resinous heart has fed." In another of the symbolically intense poems, "Vacillation," the tree is "half all glittering flame and half all green." The sap of life in the tree which transforms itself is the substance of fire and flame. Later in this verse he adds: "And half and half consume what they renew." The flame is an image of life which consumes but surprisingly rejuvenates itself, paralleling Bachelard's exposition of this image.

Bachelard recreates painterly reminiscences as well as poetic. Reading the chapter "Poetic Images of the Flame in Plant Life," in the discussion of water lilies, I relived memories of Claude Monet's garden at Giverny and visualized glorious paintings of light and flowers, concurring with Bachelard's judgment that flame and flowers naturally associate. In a detailed exposition, he expresses the reciprocal action between two image pairs which can be read forward or in reverse: "Correlatively, then, the fire flowers and the flower lights up. These two corollaries could be developed endlessly: color is an epiphany of fire, and the flower is an ontophany of light" (59). Always Bachelard stirs the imagination to surprise and wonder.

The most humanized of flames are given off by the lamp or candle, according to Bachelard. Of these bearers of the flame, he claims that the lamp is a "disciplined" flame (10). Oil lamps may seem outmoded in the twentieth century, but Bachelard revives them in terms of their evocative nature. Lamplight invokes a "fusion of imagination and memories" (8). "Where a lamp once reigned, now reigns memory" (11). From this thought image, my imagination leaps to the remembrance of

Georges de La Tour's painting, "Magdalen with the Smoking Flame," in which Mary Magdalene appears to meditate on the hereafter with a skeletal mask in her lap. Bachelard renews the linkage and nexus of images, teaching us to imagine and dream, or, in his vernacular, to find joy in reverie.

Bachelard's example of the eighteenth-century physicist of the flame trying to force two candles to burn together, "two passionate hearts trying in vain to help each other burn" (24), again projects to Yeats. "Solomon and the Witch" expresses in a delightful way the yearned-for union between two lovers: "Yet the world ends when these two things,/ Though several, are a single light,/ When oil and wick are burned as one."

Single images inevitably expand into complete cultural complexes for Bachelard. On a broader canvas Bachelard explores the Empedocles complex as an imaginative construct resulting from a particular and individual response to fire. It is named after Empedocles, the pre-Socratic philosopher who threw himself into the flames of Mount Etna. Beginning with *The Psychoanalysis of Fire* and continuing here, he will further expand the definition of this complex in an entire chapter in *A Poetics of Fire*. The moth is a "tiny Empedocles" (34), helpless to resist the allure of the flame, preferring to die in a blaze rather than live ignominiously. Certain people, those compelled by an Empedocles complex, likewise prefer dramatic life choices to mundane ones.

Bachelard illustrates how images bridge the sensate, perceived world and the interior, affective life. Empirical evidence carries emotional connections. In Chapter Three he addresses the specific quality of verticality of the flame, "The flame is an inhabited verticality," and he speaks of "the verticalizing will" (40). "Everything upright, everything vertical in the cosmos, is a flame," and "everything that rises possesses a flames's dynamics" (43). How absurd but how true!

Teilhard de Chardin, in a metaphysical vein appropriate to

the theologian, rhapsodizes about the imagery of light and flame at the heart of matter:

> Throughout my whole life, during every moment I have lived, the world has gradually been taking on light and fire for me, until it has come to envelop me in one mass of luminosity, glowing from within . . . the purple flush of matter imperceptibly fading into the gold of spirit, to be lost finally in the incandescence of a personal universe. . . . This is what I learnt from my contact with the earth—the diaphany of the divine at the heart of a glowing universe, the divine radiating from the depths of matter a-flame. ("Introduction," *Le Divine Milieu*)

Bachelard echoes these profound feelings when he relates that the "life flame of a being that flowers is a straining toward the world of pure light" (60).

The *Divine Comedy* is central to any discussion of fire imagery. In the traditional place of fire, the Inferno, Dante uses the image sparingly. The soul's ultimate state of depravity in the pit of hell is rendered by images of frozen ice, not fire. The relationship between fire and the burning away of impurities in the soul is first revealed as the massive wall of fire in Purgatory, a symbolic baptism through which all who seek salvation must pass.

Bachelard's primary focus is the exploration of spirit and flame: "The flame illustrates every form of transcendence" (41). At the base the flame is red and gold; the purgatorial one purifies itself of matter while growing toward the ultimate goal of luminosity at the summit where it becomes a divine, white light. For Bachelard, "consciousness and the flame have the same destiny in verticality" (19). Certainly this is supported by Dante's poetic vision in the *Paradiso* where in the circling spheres souls become ever lighter as they ascend to the empyrean of pure light.

To Bachelard, "fire, air, light, everything that rises also partakes of the divine" (60). Aren't we reminded in *terza rima* of the imagery Dante uses to depict the nature of the Godhead in the first canto of the *Paradiso*?

> The Glory of Him who moves all things soe'er
> Impenetrates the universe, and bright
> The splendor burns, more here and lesser there.
> (trans. Dorothy L. Sayers)

This distilled, pure fire can be compared to the special characteristic of the candle, which Bachelard pictures as a "superflame which glistens above its very tip" (46).

The *Paradiso* is suffused with this dominant image of the luminous flame. The riveting flame magnetizes Dante to accompany Beatrice's lead as he gazes into her eyes. Beatrice's eyes, "Her holy eyes aflame," provide the guiding light for the pilgrim. She compels him to follow her toward the ultimate blazing presence:

> Toward Beatrice's self I moved me, turning;
> But on mine eyes her light at first so blazed
> They could not bear the beauty and the burning.

The flame image is used as a corollary for the ascending state of the individual soul purifying itself in aspiring toward union with God whose ultimate nature is likened to a radiant flame. The image of a flame expresses, moreover, the highest heaven, the Primum Mobile, where all salvation circles around the God-head:

> Purest in flame the inmost circle proved.
> Being nearest the Pure Spark, or so I venture,
> Most clearly with Its truth it is engrooved.

This unadulterated light emerges only when a flame burns away all impurities. Bachelard calls our attention to an aspect of the flame which we have often observed but have perhaps never really seen:

> Fire receives its real existence only at the conclusion of the process of becoming light, when, through the agonies of the flame, it has been freed of all its materiality. . . . The light, then, is the true driving force determining the ascensional being of the flame.(43)

In his vision of the summit, Dante gives form more adequately than any other poet in the Western world to the ultimate human experience of the numinous. Dante's fourteenth-century choice of the flame as a supreme image is reinforced in the twentieth century by Bachelard's phenomenologically sensitive observations. The ascensional quality is a natural characteristic of the flame. Bachelard makes the point tellingly. The horizon of the earth may be responsible for beckoning mortals to explore the world and to seek distant goals, whereas the yearning to ascend the heights, the aspiring to reach the ultimate, is provoked by the contemplation of the flame of a candle.

Joanne H. Stroud
Founding Fellow,
The Dallas Institute
of Humanities and Culture

Acknowledgments

Thanks are in order to recognize with gratitude all those who valued this project and gave it loving attention at various stages. First to Scott Dupree who even from Singapore managed to stay on top of the work. Mary Bonifield captained the translation from first to last, from the minute checking on words to the most complicated electronic transfers. Marie Basalone has led many of the Fellows of the Dallas Institute into computer literacy and demonstrated how we could speed our publications schedule. Angela Fritsen and Amy Corzine were meticulous in their attention to details. Gerald Burns employed his poetic talents to proof our manuscript and supply key phrasing. Two husband and wife partnerships contributed significantly to the final product. Robert and Ki Kugelman supplied the scholarship for the footnotes and endnotes. Don and Louise Cowan are always generous with their counsel. Françoise Laye, in Paris with the Presses Universitaires de France, extended assistance whenever needed. It was indeed a communal effort.

J. H. S.

The Flame of a Candle

Preface

I

NOT OVERBURDENED by any erudition or imprisoned by a uniform mode of inquiry in this little book of simple reverie, I would like to show in a series of short chapters how reverie is renewed when one contemplates a single flame. Of all the objects in the world that invoke reverie, a flame calls forth images more readily than any other. It compels us to imagine; when one dreams before a flame, what is perceived is nothing compared to what is imagined. The flame carries its wealth of metaphors and images into the most diverse realms of meditation. Take it as the subject of one of the verbs which express life and you will see it enlivens that verb. Generalizing philosophers assert this with dogmatic calm: "What is called *Life* in creation is, in all forms and in all beings, one and the same spirit, a single flame."[1] But such a generalization reaches its objective too quickly. It is rather in the multiplicity and particularity of images that we should sense how imagined flames work on the imagination. The verb *to inflame* must enter into the psychologist's vocabulary. It governs an entire realm of the expressive world. Images of inflamed language inflame the psyche, producing a pitch of excitation that a poetic philosophy must render exactly. When flame is taken as the object of reverie, the coldest *metaphors* become *images*. Metaphors are often only displaced thoughts, the result of an attempt to say something more clearly or differently, but images—real images—as imagination's earliest form of life, leave the real world behind for an imagined, for an imaginal

---

1. Herder, cited by Albert Béguin, *L'Ame romantique et le rêve* (Marseille, Cahiers du Sud), I:113.

world. Through the imagined image we are made familiar with that absolute in the realm of reverie we know as *poetic reverie*. Correlatively, as I attempted to show in my last book (but has a book ever fully expressed all its author's convictions?), we know our dreaming being as producer of reveries. A dreamer happy in dreaming, actively participating in his reverie, has found out one of being's truths, a vista of human existence.

Of all images, images of the flame—the most artless as well as the most refined, the wisest as well as the most foolish—bear the mark of the poetic. Whoever dreams of a flame is a potential poet. All reverie in the presence of a flame is admiring reverie. Whoever dreams of a flame is in a state of primal reverie. This primordial admiration is rooted in our distant past. We have a natural, I daresay an innate admiration for the flame. The flame intensifies the pleasure of seeing beyond what is usually seen. It compels us to look.

The flame summons us to see for the first time. We have a thousand memories of it; we dream of it. It takes on the character of a very old memory, and yet we dream as everyone dreams; we remember as everyone else remembers. Then, obeying one of the *most consistent laws* of this reverie that happens before a flame, the dreamer dwells in a past which is no longer his alone, the past of the world's first fires.

II

Thus, contemplating a flame perpetuates a primordial reverie. It separates us from the world and enlarges our world as dreamers. In itself the flame is a major presence, but being close to it makes us dream of far away, too far away—"One loses oneself in reveries." The flame is there, feeble and tiny, struggling to stay in existence, and the dreamer goes on to dream of elsewhere, losing his own being by dreaming on a grand, on a too grand scale by dreaming of the world.

The flame is a world for the solitary man.

So if the dreamer of the flame speaks to it he speaks to himself and is suddenly a poet. By enlarging the world and its destiny, by meditating on the destiny of the flame, the dreamer enlarges language because he expresses some of the world's beauty. Through such pancalizing expression the psyche itself is elevated and enlarged. Meditation on a flame gives the dreamer's psyche vertical nourishment, verticalizing food. Aerial nourishment as opposed to "earthly food"—there is no more active principle for giving poetic specificity an energetic direction. I shall return to poetic specificity in a special chapter to illustrate the advice every flame offers: burn high, ever higher, to be sure you will give off light.

If one is to reach this "psychic height" every impression must be filled with poetic matter. The poetic contribution is enough, I think, to make me hopeful of unifying these reveries gathered under the sign of the candle. This monograph could be subtitled *The Poetry of Flames*. Indeed, in my desire to pursue a single line of reverie I have separated it from a more general monograph that I expect to publish under the title *A Poetics of Fire*.

III

At present, by limiting the scope of my inquiries to a single example, I hope to arrive at a *concrete aesthetic*, one not belabored by philosophical polemics or rationalized through facile generalizations. The flame and only the flame can give concrete being to all its images, to all its phantoms.

So simple an object—a flame!—will be invested with literary images which I hope will define a communion of imaginations. Through these literary images of the flame, surrealism has some guarantee of being rooted in reality! The most fantastic images of the flame converge. By a special privilege they become real images.

The paradox of our inquiries into the literary imagination—to find reality through spoken words, to sketch in written words—may yet be resolved. *Spoken images* are a translation of the extraordinary excitation of our imagination by the simplest flame.

IV

I must explain yet another paradox. In wishing to experience literary images by bringing them fully into the present, along with the even greater ambition of proving that poetry is an active force in life today, is it not uselessly paradoxical to situate so many reveries under the sign of the candle? The world goes quickly; the century accelerates. Ours is no longer the era of tapers and candlesticks. Only old-fashioned dreams are still associated with such obsolete things.

There is an easy answer to these objections: dreams and reveries are not modernized as rapidly as our actions. Our reveries are true, deeply rooted *psychic habits*. Active life does not disturb them in the least. The psychologist has an interest in finding all paths into the most ancient familiar things.

Reveries of this faint light will lead us back to the wee space of familiarity. It seems that there are dark corners in us that tolerate only a flickering light. A sensitive heart loves fragile values. It communes with values that struggle, hence with the feeble light that struggles against darkness. Thus all our reveries of *faint light* retain psychological reality in today's life. They have a meaning and, I would boldly state, a function. In fact they may give a psychology of the unconscious a complete image system for exploring dreaming existence subtly and naturally, without ambiguity. Reveries of faint light make the dreamer feel at home; the dreamer's unconscious becomes home for him. The dreamer, that twin of our being, that chiaroscuro of the thinking person, feels secure in his existence during this reverie in faint light. Whoever trusts in the reveries of faint light will discover this psychological truth: the tranquil unconscious, an unconscious without nightmares and in harmony with its reverie, is quite precisely the psyche's chiaroscuro, or better yet, the chiaroscuro's psyche. Images from this faint light teach us to love this chiaroscuro of innermost vision. As soon as he begins to love his reverie, a dreamer who wishes to know himself as a dreaming being far from the clarities of thought is tempted to formulate an aesthetics for this chiaroscuro of the psyche.

A lamplight dreamer will understand instinctively that images of faint illumination are our innermost pilot lights. Their glimmers become invisible when thought is at work, in the clarity of consciousness. But when thought is at rest these images remain active.

Consciousness of the chiaroscuro of consciousness is present in such an enduring way that every being awaits an awakening, an awakening of being. Jean Wahl knows this. He says it in a single line:

> O little light, O source, delicate dawn.[2]

V

I propose, therefore, to transfer the aesthetic values of the painter's chiaroscuro to the realm of the psyche's aesthetic values. If I am successful, I will have cleared away what is reductive and pejorative in the notion of the unconscious. The shadows of the unconscious so often exploit this world of glimmers in which reverie finds a thousand delights! George Sand had an intimation of this transition from the world of painting to the world of psychology. In a footnote to the text of *Consuelo* she evokes chiaroscuro:

> I have often tried to determine in what that beauty consisted, and how it would be possible for me to *describe* it[3] if I wished to disclose the secret to another mind. "What!" someone will say, "do you mean that external objects, without color or shape, in disorder and unlighted, can take on an aspect that appeals to *the*

2. Jean Wahl, *Poèmes de circonstance*, Éd. Confluences, 33.
3. The italics are my own.

eyes and the mind?" None but a painter could ever say to me: "Yes, I understand." He would recall Rembrandt's *Philosopher in His Study*: that enormous room, three-fourths in darkness, those endless stairways which wind no one knows how; those vague lights [which blaze up and go out, you know not why, in different parts] of the picture; that whole scene, indefinite yet clear; that powerful coloring, which after all is only light brown and dark brown; that magical use of chiaroscuro, that play of light and shadow on the most trivial objects, a chair, a jug, a copper urn; and lo! those objects which do not deserve to be glanced at, much less to be painted, become so interesting, so beautiful after their manner, that you cannot take your eyes from them. They have received the breath of life, they exist.[4]

George Sand sees and states the problem, which is not how to paint this chiaroscuro—for that is the privilege of great artists—but how to "describe" it, how to transcribe it. I myself would go further: how does one *inscribe* this chiaroscuro into the psyche right up to the frontier between a deep brown and a lighter brown psyche?

In fact this is a problem that has tormented me for the twenty years that I have written books on Reverie. I do not know how to express it better than George Sand has done in her short note. In a word the chiaroscuro of the psyche is reverie, a calm and calming reverie, faithful to its center, illuminated at the center, not crowding its contents but always overflowing a little, impregnating its penumbra with light. One sees clearly into himself, yet one dreams. One does not risk all of one's light. One is not the plaything or the victim of this musing that comes by night and delivers us, bound hand and foot, to the despoilers of the psyche, to those brigands that haunt the forests of nocturnal sleep which are our dramatic nightmares.

The poetic aspect of a reverie makes us yield to this golden state of the psyche that keeps consciousness awake. Reveries

---

4. *Consuelo* (Boston and New York, 1900-1902), III:270.

in the presence of a candle will make paintings. The flame will maintain us in that awareness of reverie that keeps us awake. One falls asleep before a fire but not before the flame of a candle.

VI

In a recent book I attempted to establish a radical difference between reverie and nocturnal dream. In the nocturnal dream there reigns a phantasmal illumination. Everything appears in a false light. There one often sees too clearly. Mysteries themselves are drawn in bold lines. The scenes are so distinct that nocturnal dreams turn easily into literature—literature, but never poetry. All fantastic literature finds schemas in the nocturnal dream upon which the animus of the writer works. It is by way of the animus that the psychoanalyst studies dream images. For him the image is double, always signifying something other than itself. This is a caricature of the psyche. One racks the brain to find the real within the caricature. To rack the brain, to think, always to think. To enjoy images, to love them for themselves, would demand that the psychoanalyst accept a poetic education on the fringes of all such erudition. Hence fewer dreams by way of the animus and more by way of the anima. Less of understanding by way of intersubjective psychology and more of sensitivity through a psychology of the familiar.

From the point of view I am adopting in this little book, reveries on familiar things eschew drama. A notion of the fantastic orchestrated by concepts taken from nightmare experiences will not hold our attention. When I encounter an image of the flame too singular to make my own, to place in the chiaroscuro of personal reverie, I will at least avoid long commentaries. In writing about the candle I want to reach the gentle realms of the soul. One must want vengeance in order to imagine hell. In nightmare beings there is a hellfire complex that I do not want to feed, whether from near or from afar.

To sum up, studying the existence of a reverie-dreamer by means of the images of faint light, images that have been human ones for a very long time, guarantees the homogeneity of a psychological inquiry. There is a relationship between the burning pilot light and the soul that dreams. Time is as slow for one as for the other. The same patience appears in the dream and in the glimmer of light. Time is deepened; images and memories reunite. He who dreams about a flame unites what he sees and what he has seen. He recognizes the fusion of imagination and memory. He thus becomes open to all the adventures of reverie; he accepts the help of great dreamers; he enters the world of the poets. From that moment on, reverie inspired by a flame, so much a unit in its origin, becomes an abundant multiplicity.

To put some order into this multiplicity, I will provide a quick commentary on the sometimes very diverse chapters of this simple monograph.

VII

The first chapter is still a prefatory one. I must explain how I resisted the temptation to write an erudite volume about flames. It would have been a long but easy one to write. It would have been enough to have made it into a history of the theories of light. From century to century the problem has been taken up again and again. But some of the greatest minds who worked with the physics of fire were never able to give scientific objectivity to their work. The history of combustion remains, until Lavoisier, a history of pre-scientific views. The examination of such doctrines belongs to a psychoanalysis of objective knowledge. This psychoanalysis would have to get rid of images to determine how ideas are organized.[5]

---

5. Cf. Gaston Bachelard, *La Formation de l'esprit scientifique. Contribution à une psychanalyse de la connaissance objective* (Paris: Vrin, 1938).

The second chapter is a contribution to a study of solitude, to an ontology of solitary existence. The isolated flame is the witness of solitude, a solitude that unites flame with dreamer. Thanks to the flame the solitude of the dreamer is not a solitude of the void. Solitude, through the grace of this little light, has become concrete. The flame illustrates the solitude of the dreamer; it illuminates the pensive brow. The candle is the polestar of the blank page. I will bring together several texts taken from poets as commentary on this solitude. I welcome these texts so readily that I am fairly sure they will be well received by the reader. Thus I acknowledge my conviction about images. I believe that the flame of a candle is, for many dreamers, an image of solitude.

Though I have been careful to avoid detours into pseudoscientific research, I was often attracted to fragmentary thoughts which do not prove but rather give unequaled impetus to reverie through pithy assertions. Then it is not science but philosophy that dreams. I have read and reread the work of Novalis. From him I have learned great lessons in how to meditate upon the verticality of the flame.

When I studied the technique of the waking dream in one of my first books on the imagination,[6] I noted the invitation to a dream of flight that I received from a dawning universe, a universe which bears light at its uppermost heights. I then analyzed the psychoanalytic technique of the waking dream instituted by Robert Desoille. It involved using the suggestion of pleasant images to lighten the existence of a person weighed down by his faults, put to sleep by his boredom with life. As images evolved, the guide evolved for the patient into a guide to evolution. This guide proposed an imaginary ascent, an ascent which had to be illustrated by carefully ordered images, each having an ascensional property. The guide

6. *Air and Dreams: An Essay on the Imagination of Movement* (Dallas, 1988).

nourished the oneirism of the dreamer, offering images at the proper point to set the rising psyche into motion again and again. This rising psyche is of no value unless it rises higher, ever higher. The images of this psychoanalysis through elevation must systematically become too high in order to be certain that the patient will be in the midst of a fully metaphorical life when he leaves the lower levels of being.

But perhaps only the solitary flame can be an ascensional guide for the meditating dreamer. It is a model of verticality.

Numerous poetic texts will help us exploit this verticality, in light and by means of light, where Novalis dwelled while meditating upon the upright flame.

After examining some philosophers' dreams, I returned in the fourth chapter to the problems that we know so well, the problems of the literary imagination. A large book would not suffice to study the flame and pursue all the metaphors it suggests throughout literature. One might wonder if the image of the flame could not be linked with every slightly brilliant, or even would-be brilliant, image. The result would be a book on general literary aesthetics, created by classifying all those images that could be heightened by putting an imaginary flame within them. Such a work, which would show that the imagination is a flame, the flame of the psyche, would be very pleasant to write. One could spend his entire life on it.

In speaking of trees and flowers, I was able to say how the poets bring them to life, to abundant poetic life through the image of flames.

From the candle to the lamp there is for the flame something like a conquest of wisdom. The flame of a lamp, thanks to man's ingenuity, is now disciplined. It is given over completely to its task, both simple and lofty, as giver of light.

I wish to conclude my work by meditating upon this humanized flame. It would require an entire book to do justice to the transition from the cosmology of the flame to the cosmology of light. Since I shall not cover such a broad subject, I want this monograph to remain within the homogeneity of reveries about this faint light and dream in that innermost realm where lamp and candlestick are united, the ancient dwelling place of that indispensable pair to which we continually return for dreams and remembrances.

I have found a great aid to reverie in the work of a master who is familiar with the dreams of memory. In many of Henri Bosco's novels, the lamp is a character in every sense of the term. The lamp has a psychological role to play in the psychology of the house and in the psychology of family members. When the absence of an important person creates an emptiness in the home, Bosco's lamp—coming from who knows what in his past—maintains its presence and awaits the exiled one with a lamp's patience. Bosco's lamp keeps all the memories of family life alive, all the memories of a childhood, the memories of every childhood. The writer writes for himself, and he writes for us. The lamp is the spirit that watches over his room, over every room. It is the center of a dwelling, of every dwelling. One can no more conceive of a house without a lamp than a lamp without a house.

Meditation upon the familial being of the lamp will thus allow us to take up once more our reveries on the poetics of innermost spaces. Again we find all the themes that were developed in *The Poetics of Space*. With the lamp we come back to the refuge of evening reverie in ancient dwelling places, dwelling places now lost but faithfully inhabited in our dreams.

Where a lamp once reigned, now reigns memory.

Finally, in order to leave a slightly personal mark in this little book on the reveries of others, I thought I could add an epilogue to invoke the solitude of work, the vigils during which, far from relaxing into facile reveries, I worked persistently, believing that with the work of thinking one enlarges the mind.

1

Candles and Their Past

> Flame, winged commotion,
> O breath, red glint of the sky
> —whoever could unravel your mystery
> would know about life and death . . .
> MARTIN KAUBISH
> Anthologie de la poésie allemande

I

LONG AGO, in a long ago even dreams themselves have forgotten, the flame of a candle made wise men think; it provided the solitary philosopher with a thousand dreams. On his table, next to objects imprisoned in their shapes, next to those books that teach one so slowly, the flame of the candle summoned endless thoughts and aroused immeasurable images. For a dreamer of worlds at that time, the flame was among the world's phenomena. The system of the world was studied in large books, and now a simple flame—oh mockery of erudition!—insists upon its own enigma. Is not the world alive in a flame? Does it not have a life of its own? Is it not the visible sign of some innermost being, the sign of some secret power? Does it not hold within itself all the internal contradictions that make an elementary metaphysics dynamic? Why search the dialectics of ideas when the dialectics of fact and being exist at the heart of a simple phenomenon? The flame is a being lacking substance, but for all that it is strong.

What a field of metaphors we would have to examine if we wanted to unfold the images that unite flame and life and write a "psychology" of flames alongside a "physics" of the fires of life! Mere metaphors? In that time of distant knowledge when the flame made wise men think, metaphors were a form of thinking.

II

But if the knowledge in old books has perished, the interest in reverie remains. In this little book we shall try to put all our documents, whether they come from philosophers or poets, into a setting of *primal reverie*. When we rediscover the roots of simplicity in our dreams or in others' of which we have heard, all belongs to us; all is intended for us. In the presence of a flame we communicate morally with the world. Even in a simple vigil the flame of a candle is the model of a tranquil and delicate life. The slightest breath will certainly disturb it, just as an alien thought will disrupt the meditation of a meditating philosopher. But when the reign of full solitude truly arrives, when the hour of tranquility truly sounds, then the same peace resides in the heart of the dreamer as in the heart of the flame and then the flame preserves its form and rises directly, like a resolute thought, towards its vertical destiny.

Hence in those times when one dreamed by thinking and thought by dreaming, the flame of the candle could be a sensitive gauge of the soul's tranquility, the measure of a refined calm that reaches into life's smallest details, bestowing gracious continuance to time spent in peaceful reverie.

Do you want to be calm? Breathe softly in front of the delicate flame as it calmly performs its work of illuminating.

III

Living reveries can thus be fashioned from very old knowledge. We will not, however, search out texts from old books of magic spells. Instead we prefer to restore to every image that we retain its oneiric layers, its haze of indefiniteness, so as to bring it into our own reverie. Only through reverie can unusual images be communicated. The intellect is an awkward analyst of untutored reveries. In just a few pages of this little essay, we will invoke texts in which familiar images are expanded into revelations of the world's secrets. With what ease the dreamer of the world moves from his bit of candle to the great luminaries of the heavens! When we are seized by such expansions in our reading, we become enthusiastic. But we can no longer bring our enthusiasm into order. From all our investigations we retain only spurts of images.

When a particular image takes on cosmic value it functions as a vertiginous thought. Such an image-thought or thought-image has no need of context. The flame seen by a seer is a phantom reality that calls for a spoken declaration. In what follows I shall give many examples of thought-images set forth in striking phrases. Sometimes such image-thought-phrases give sudden color to quiet prose. Joubert, reasonable Joubert, writes: "The flame is a moist fire."[1] Later I shall provide variations on this theme of the conjunction of flame and stream. I mention it in this preamble of a chapter to show right away how insistent are those reveries that revel in disturbing sleepy erudition. A single inconsistency is enough for them to perturb nature and liberate the dreamer from banal opinions about familiar phenomena.

Thus the reader of Joubert's *Pensées* enjoys imagining as well. He sees this moist flame, this burning liquid, flowing upward toward the sky like a vertical stream.

In passing we should note a nuance that properly belongs to the philosophy of literary imagination. An image-thought-phrase like Joubert's is a linguistic feat. In it speech surpasses thought. And the reverie that speaks is itself surpassed by the *reverie that writes*. One dares not vocalize this reverie of a "moist fire" but it can be written down. The flame was a writer's temptation and Joubert could not resist it. Rationalists must pardon those who harken to the demons of the inkwell.

Had Joubert's formulation been a thought, it would have been merely an overfacile paradox. Had it been an image, it

---

1. Joseph Joubert, *Pensées*, 8th edition, 1862, 163. The first blowlamps were sometimes called "fountains of fire." Cf. Edouard Foucaud, *Les Artisans illustres* (Paris, 1841), 263.

would have been fleeting and ephemeral. But occurring in the book of a great moralist, his formulation opens up for us the field of *serious reverie*. The tone of fantasy blended with reality gives me, simple reader that I am, the right to *dream seriously*, as though in such reveries my mind worked in all its clarity. In the serious reverie that I draw from Joubert's work, a phenomenon of the world is expressed and hence surpassed. It is expressed as transcending its own reality. It exchanges its reality for a human reality.

Refashioning for ourselves some of the images of the philosopher's cell, we may see a taper and an hourglass—two things that tell human time—on the same table but how different in style they are! The flame is an hourglass that flows upward. Lighter than falling sand, the flame molds its own shape, as though time itself always has something to do.

In peaceful meditation the flame and the hourglass express the communion of buoyant and heavy time. In my reverie they speak the communion of anima and animus time. I would love to dream of time, of the hours that flow and those that fly, if only I could combine candle and hourglass in my imagined cell.

But for the wise man that I imagine, the lesson of the flame is greater than the lesson of falling sand. The flame calls the vigilkeeper to raise his eyes from his folio, to quit his working time, his reading, his thinking time. Within the flame, even time holds its vigil.

Yes, he who keeps a vigil before the flame no longer reads. He thinks of life. He thinks of death. The flame is precarious and courageous. This light is destroyed by a breath, relit with a spark, easy birth and easy death. Life and death are well juxtaposed here. In this image of them, life and death are well-conceived contraries. The philosophers' thought-games, prosecuting their dialectics of being and nothingness in a simple logical style, become dramatically concrete before the light that is born and dies.

But when one dreams more deeply, this lovely equilibrium of life and death in our thoughts is no more. In the candle dreamer's heart the word *extinguish* has such echoes! Doubtless, words forsake their origins and take on a strange new life, a life borrowed randomly from simple comparisons. Which is the greater subject of the verb *to extinguish*, life or the candle? Metaphorizing verbs make the most heterogeneous subjects spring to action. The verb *to extinguish* can make anything die, a noise as well as a heart, love as well as anger. But whoever wants its real meaning, its original meaning, should recall the dying of a candle. Mythologists have taught us to read the dramas of light in the spectacles of the sky. But in the dreamer's cell, familiar objects become myths of the universe. A candle extinguished is a sun that dies. The candle dies even more gently than a star of the sky. The wick bends; the wick blackens. The flame has swallowed its opium from the shadow that embraces it. And the flame dies a good death; it dies in its sleep.

Every candle dreamer, everyone who dreams of a small flame knows this. Everything is dramatic in the life of things and in the life of the universe. One dreams twice when he dreams in the company of his candle. Meditation before the flame becomes, in Paracelsus's phrase, an exaltation of two worlds, an *exaltatio utriusque mundi*.[2]

Simple philosopher of literary expression that I am, I shall restrict myself subsequently to poets for evidence of this double exaltation of two worlds. The times when such dreams, such unbounded dreams, could be assisted by thoughts, by my own labored ones or those of others, are over as I have said in my opening pages.

But has poetry ever really been made out of thoughts?

---

2. Cited by Jung in *Paracelsica: Zwei Vorlesungen über den Arzt und Philoshophen Theophrastus* (Zurich, 1942), 123.

IV

To justify limiting myself to texts that lead us to serious reveries which approximate the dreams of a poet, I am going to comment on one example among many others. It comes from an aggregate of images and ideas borrowed from an old book whose ideas, no more than its images, cannot tempt us to participate in them. Separated from their historical context, the passage I shall cite can hardly be claimed as an achievement of the fantasy. Nor is it much like systematic erudition. It is only a blend of pretentious thoughts and simplistic images, quite the contrary of that excitement with images which we enjoy experiencing. It is something like a *blunder of the imagination.*

After having commented on this massive document, we will return to more refined images that have not been so crudely assembled into a system. There we will rediscover impulses that we may pursue in a personal fashion by reliving through them the joys of imagining.

V

Blaise de Vigenère, in his *Treatise on Fire and Salt* writes in his commentary on the *Zohar*:

> There is a double fire, a stronger one that devours the other. Let whoever wants to know it contemplate the flame that separates and rises from a lighted fire or a lamp or torch, because it never rises unless it is incorporated with some corruptible substance and united with air. But in this flame that rises are two flames: the one which is white shines and illuminates, having its blue source at the top, and the other which is red is attached to the wood and to the taper that it burns. The white rises directly into the heights, and the red dwells firmly below without leaving the matter that administers whatever causes the other to blaze and give off light.[3]

---

3. Blaise de Vigenère, *Traité du feu et du sel* (Paris, 1628), 108.

Here begins the dialectic of passive and active, of the moved and the moving, of the burned and the burning—the dialectic of past and present participles that satisfies philosophers in all ages.

But for a flame thinker like Vigenère, the facts must open up a horizon of *values*. The value to be conquered here is light. Light is thus a valorization of fire, a hypervalorization in that it gives meaning and value to facts that we first take to be insignificant. Illumination is truly a conquest. Vigenère in fact makes us feel the difficulty the crude flame has in becoming the white flame, in achieving the dominant value of whiteness. This white flame is "always the same, without change or variation like the other, which sometimes blackens, then becomes red, yellow, indigo, bluish-green, azure."

So the yellowish flame will be the *antivalue* of the white flame. The flame of the candle is the arena for a struggle between value and antivalue. The white flame must "exterminate and destroy" the crudeness that nourishes it. Thus for a prescientific author, the flame has a positive role in the economy of the world. It is instrumental in improving the cosmos.

The moral lesson is a foregone conclusion: moral consciousness must become a white flame by "burning the iniquities it harbors."

What burns well burns high. Consciousness and the flame have the same destiny in verticality. The simple flame of a candle points clearly to its destiny; it "goes deliberately upward, and returns to its proper dwelling place, after having accomplished its task below without changing its glimmer to any color other than white."

Vigenère's text is long. I have abridged much of it. It can be tedious. It must be tedious if it is to be considered as an *idea text* for organizing knowledge. At least as a *reverie text*, it seems to me clear evidence of a reverie that overflows all boundaries and encompasses all experiences, whether coming from man or from the world. The phenomena of the world, as soon as they acquire a little consistency and unity, turn into human truths. The *moral* that concludes Vigenère's text should surge back over the entire story. This moral was latent in the dreamer's interest in his candle. *He looked at it morally.* It was for him a moral access to the world, an access to the morality of the world. Would he have dared to write of it if he had seen no more than burnt tallow? The dreamer had on his table what we may call a *model-phenomenon*. The most vulgar material of all produces light. It purifies itself in the very act of giving off light. What an eminent model of active purification! The impurities themselves, by being annihilated, provide pure light. Evil is thus the nourisher of good. In the flame, the philosopher encounters a *model-phenomenon*, a cosmic phenomenon, a model of humanization. In imitating this model-phenomenon we "burn up our iniquities."

The purifying and purified flame illuminates the dreamer twice, through the eyes and through the soul. Here metaphors are realities, and reality, because it is *contemplated*, is a metaphor of human dignity. One contemplates reality by metaphorizing it. The value of Vigenère's document would be distorted if analyzed from a symbolic perspective. The image demonstrates; the symbol affirms. The naïvely contemplated phenomenon is not, like the symbol, charged with history. The symbol is a conjunction of traditions with multiple origins. Not all of these origins are reanimated by contemplation. A culture's present is stronger than its past. Vigenère's study of the *Zohar* did not prevent him from grasping, in all the primitiveness of his reverie, something that was supposed to be knowledge from an old book. As soon as one is invited by his reading to dream, one's reading stops. If the candle illuminates an old book that speaks of flames, the ambiguity of thoughts and reveries becomes extreme.

There is no symbol or double language here that might translate matter into spirit or vice versa. With Vigenère we are within the powerful unity of a reverie that unites man and his world, the powerful unity of a reverie that cannot be divided into a dialectic of subjective and objective. In such a reverie the world assumes a human destiny in all of its objects. Now in the innermost recesses of its mystery the world desires this destiny of achieving purification. The world is the seed of a better world, as man is the seed of a better man, as the heavy yellow flame is the seed of a pale, white flame. In returning to its natural place through its *whiteness* and the dynamism of its conquest of whiteness, the flame does not merely obey Aristotelian philosophy. A value greater than all those which rule physical phenomena has been won. This returning of things to their natural places is an ordering, certainly, a restitution of order in the cosmos. But, in the case of the white light, a moral order takes precedence over the physical order. The natural environment toward which the flame tends is a moral environment.

That is why the flame and its images show the values of man to be values of the world. They unite the morality of the "little world" with the majestic morality of the universe.

Over the centuries mystics who speak of the *volcano's finality* have said nothing more than this when they claim that the earth "purges itself of its impurities" through the beneficent action of its volcanoes. Michelet repeated the same thing again in the last century. If man can think so grandly, he can certainly dream on a small scale as well and believe that his little light may serve to purify the world.

VI

Of course if we were to direct our inquiries to liturgical problems or lean on the type of major symbolism originally formed through moral and religious values, it would not be difficult to find more dramatic symbols of the flame and the torch—the masculine name for a gloriously burning flame—than those symbols born naïvely in the candle dreamer's reveries. But I believe that there is some value in pursuing a reverie that welcomes the most far-fetched comparisons amid familiar phenomena. A comparison is often the beginning of a symbol, one which does not yet bear its full responsibility. The disequilibrium between what is perceived and what is imagined quickly reaches its limits. The flame is no longer an *object of perception*. It has become a *philosophical* object. Then anything is possible. In the presence of his candle the philosopher can easily imagine that he is the witness of a world aflame. For him the flame is a world moving toward becoming. In it the dreamer sees his own being and his own becoming. Space moves in the flame; time is active. Everything trembles when the light trembles. Is not the becoming of fire the most dramatic and the most alive of all becomings? The world moves rapidly if it is imagined on fire. Hence the philosopher can dream everything—violence and peace—when he dreams of the world before his candle.

2

The Solitude of the Candle Dreamer

> Already my solitude is prepared
> to burn him who burns it.
> LOUIS EMIÉ, *Le nom du feu*

I

Now, AFTER A brief preliminary chapter in which I sketched out the research topics that an historian of ideas and experiences would need to pursue, I return to my simple task as a seeker of images, of images appealing enough to stabilize our reveries. The flame of the candle summons reveries from memory. It provides us occasions, in our distant memories, for solitary vigils.

But does the solitary flame alone intensify the dreamer's solitude or comfort his reverie? Lichtenberg said that man so needs companionship that in his solitary dreaming he feels less forlorn before a lighted candle. Albert Béguin was so struck by this thought that he entitled the chapter dedicated to George Lichtenberg, "The Lighted Candle."[1]

But every "object" that becomes an "object of reverie" takes on a peculiar character. How I would like to undertake the immense labor of bringing together a museum of "oneiric objects," objects that are made oneiric by a familiar reverie on familiar objects. Everything in a house would thereby have its "double," not a nightmare phantom but a kind of ghost that haunts the memory, that revivifies remembrance.

---

1. Béguin, *L'Ame romantique et le rêve*, I:28.

An oneiric personality, then, characterizes each important object. The solitary flame has a character different from that of the fire in the hearth. The fire in the hearth may distract the keeper of the flame. A man who stands before a talkative fire can help the wood to burn; he places an extra log at the right moment. The man who knows how to keep warm supervises the Promethean deed. He acts out his little Promethean gestures and so is proud to be perfect keeper of the flame.

But the candle burns alone. It has no need of a servant. We no longer have candle-snuffers and snuffer stands on our tables. For me the time of candles is the time of the "bougies à trous." Tears, hidden tears, flow along its lachrymal canals, lovely example for a grieving philosopher to imitate! Already Stendhal knew how to recognize good candles. In his *Memoirs of a Tourist* he tells of going to the best grocer in the area to replace the innkeeper's dirty candle-ends with good wax candles.

And so we should return to our dreams—when we are all alone—by recalling these good wax candles. The flame is alone, naturally solitary; it wants to remain alone. At the end of the eighteenth century a physicist investigated flames by attempting in vain to join the flames of two candles. He placed the candles with their wicks side by side. But the two solitary flames, in their rapture of increasing and rising, failed to unite, and each kept its own vertical energy, maintaining at the top its own delicate tip. What a symbolic disaster this physicist's "experiment" is—emblem of two passionate hearts trying in vain to help each other burn!

For a dreamer the flame should at least be the symbol of a being absorbed in his own becoming! The flame is a becoming-being, a being-becoming. If we could feel the flame alone and intact, the flame in the drama of being-becoming, destroying itself by shedding light—these thoughts well up from the images of a great poet. Jean de Bosschère writes:

> My thoughts, in the fire, lost the tunics
> by which I knew them;
> they are consumed in the blaze
> of which I am the origin and the sustenance.
> And yet I am no more.
> I am the interior, the pivot of the flames.
> And yet I am no more.²

To be the pivot of a flame! What a great and powerful image of dynamism in all its unity! Jean de Bosschère's flames, the flames of Satan the Obscure, never tremble. They may be considered the emblem of a grand enterprise.

II

With Jean de Bosschère a vital heroism models itself on an energetic flame that "rends its tunics." But there are flames of a more peaceful solitude. They speak more simply to lonely consciousness. In just five words another poet speaks the axiom of solace in these two solitudes:

> Single flame, I am alone.³

Sadness or resignation? Sympathy or despair? What is the tone of this appeal for impossible communication?

To burn alone, to dream alone—a great symbol, a twofold incomprehensible symbol. On the one hand it indicates the woman who, all afire, must remain alone without saying a word, and on the other the taciturn man who has only solitude to offer.

But the solitude of one able to love and be loved—what a jewel! The novelists tell us of the sentimental beauties of these secret loves, these undeclared flames. What a novel could be written if one could continue the dialogue begun by Tzara:

> Single flame, I am alone,

---

2. Jean de Bosschère, *Derniers poèmes de l'Obscure* (Paris, 1948), 148.
3. Tristan Tzara, *Où boivent les loups*, 15.

but does this dialogue not continue in silence, in the silence of two solitaries?

But one must speak while dreaming. In the evening reverie before his candle, the dreamer devours the past, indulges in a false past. The dreamer dreams of what he could have been. In rebellion against himself, he dreams of what he should have been and of what he should have done.

In the undulations of reverie this rebellion against oneself is quieted. The dreamer gives himself up to the melancholy of reverie, a melancholy that blends actual memories with the memories of reverie. It is in this blending, again, that one becomes sensitive to the reveries of others. The candle dreamer communicates with the great dreamers of a *former life*, with solitary life's great reservoir.

III

If my book could be what I wanted it to be, if by reading poets I could assemble enough adventures in reverie to break down the barrier that keeps us out of the Poet's Kingdom, I would like to find a truly conclusive image at the end of every paragraph and after every lengthy sequence of images, one that would seem overwrought to the reasoning mind. Aided by the imagination of others, my reverie would thus go beyond my own dreams.

In this short paragraph, to tell how one might go beyond memories of solitude and memories of woe as well, I shall evoke by candlelight a literary document in which Théodore de Banville speaks of Camoens' vigil. When a poet speaks *sympathetically* of another poet, what he says is doubly true.

Banville reports that when Camoens' candle would go out, the poet would continue to write his poem by the light of his cat's eyes.[4]

By the light of his cat's eyes! One must believe in such a

---

4. Théodore de Banville, *Contes bourgeois*, 194.

gentle, delicate light as though in something far beyond any mundane light. The candle is no more, but it once existed. It *began* its vigil while the poet began his poem. The candle led a communal life, an inspired life, inspiriting life along with the inspired poet. Line after line, in the fire of inspiration the poem unfolded its own ardent life by candlelight. Every object on the table had its glimmer of light. And the cat was there, sitting on the poet's table, its tail white against the inkstand. It was gazing at its master, at its master's hand moving over the paper. Both candle and cat were watching the poet with his fiery gaze. The gaze was everything in that little universe of the lighted table in the worker's solitude. How then could everything else not maintain the fervor of his gaze, of his light? A decrease in one is offset by increased cooperation from the others.

Moreover, weak beings move beyond themselves more delicately and less violently than strong ones. The solitude of the candle is smoothly continued in the solitude of the non-candle. Every object in the world, loved for its own sake, has a right to its own nothingness. Every being pours out being, a little of its being, the shadow of its being, into its own non-being.

Hence, in the subtle harmony that a philosopher of ultra-dreams hears between beings and non-beings, the being of the cat's eye can assist the non-being of the candle. The spectacle of a Camoens writing at night was just that great! Such a spectacle has its own kind of temporality. The poem itself wants to reach its end; the poet wants to attain his goal. The moment the candle falters, how can one fail to see that the eye of the cat is a vessel of light? Certainly Camoens' cat was not startled when the candle died.[5] The cat, that vigilant animal, that attentive being which watches while sleeping,

---

5. Let me point out that the cat is not a timorous being. We believe all too easily that what is weak is fragile. Le Sieur de La Chambre thinks, for example, that when the glow-worm is afraid, his light goes out. Cf. Le Sieur de La Chambre, *Nouvelles pensées sur les causes de la lumière* (Paris, 1634), 60.

maintains its vigil with a light that is in keeping with the face of a poet illuminated by genius.

IV

Now that an overwrought image has made us aware of the dramas of this faint light, we need no longer favor insistently visual images. Dreaming by candlelight, idle and solitary, one soon learns that this life which glows also speaks. Here again the poets will teach us to listen.

The flame whispers and whimpers. The flame is a suffering being. Somber murmurings emerge from this Gehenna. Every little pain is a sign of the world's pain. A dreamer who has read the works of Franz von Baader finds muffled, miniature bursts of lightning in the cries of his candle. He hears the noise of a being that is burning, that *Schrack* that Eugene Susini tells us is untranslatable from German into French.[6] Strange to say, the phenomena of sound and sonority are the least translatable elements from one language to another. The acoustic space of a language has its own reverberations.

But do we know how to welcome into our mother tongue the distant echoes that reverberate at the hollow center of words? When reading words we see them; we no longer hear them. What a revelation the good Nodier's *Dictionary of French Onomatopoeia* was for me! It taught me to explore with my ear those cavities in syllables that constitute the sonic superstructure of a word. In amazement and wonder I learned that to Nodier's ear, the verb *clignoter* [to wink, to blink, to flicker] was an onomatopoeia for the flame of the candle! Certainly the eye moves; the lid trembles when the flame trembles. But an ear entirely given over to an awareness of its own hearing has already heard the uneasiness of the light. You are dreaming and no longer looking. And here the brooklike flow of the flame's

---

6. Eugène Susini, *Franz von Baader et la connaissance mystique* (Paris: Vrin), 321.

CHAPTER TWO • 29

noises is thwarted; the syllables of the flame coagulate. Listen carefully—the flame flickers. Primary words must imitate what is heard before translating what is seen. The three syllables of the flickering candle flame collide, break against one another. *Cli-gno-ter*: not one syllable will blend into the other. The uneasiness of the flame is registered in the slight hostilities of the three sounds. A dreamer of words never fails to sympathize with this sonorous drama. *Clignoter* is one of the most tremulous words in the French language.

Ah! these reveries go too far! They can only emerge from the pen of a philosopher lost in his dreams. He forgets the modern world where *clignotement* [blinking] is a sign studied by psychiatrists, where a *clignotant* [blinker] is a mechanical device that obeys the touch of the motorist's finger. But words, by allowing themselves to be so many things, lose their particularity. They forget the first thing, the most familiar thing, the most basic of familiar things. A candle dreamer, a dreamer who remembers having been companion to this faint light, relearns primal simplicity by reading Nodier.

As I indicated in my preliminary chapter, a flame dreamer easily becomes a flame thinker. He wants to understand why the silent being of the candle suddenly groans. For Franz von Baader this crackling, this *Schrack*, "precedes every inflammation, whether it be noisy or silent." It is produced "by the encounter of two opposing principles in which one represses or subordinates the other to itself." Always while burning the flame must be re-inflamed to sustain the authority of its light in opposition to gross matter. If we had a more sensitive ear we would hear all the echoes of these intimate agitations. Sight synthesizes at too little cost. The rustlings of the flame cannot be so readily summed up. The flame speaks of all the struggles it must bear with to maintain its unity.

But more troubled hearts are not soothed by cosmological views that consign the misfortunes of a thing to a general Gehenna. For a flame dreamer the lamp is a companion of his states of soul. If it trembles, it presages an uneasiness that will disturb the whole room. The moment the flame twitches, so does the blood in the dreamer's heart. The flame is in anguish, and breath moves through the dreamer's throat in fits and starts. A dreamer, physically at one with the life of things, dramatizes the insignificant. In this particulate reverie everything has human meaning for this dreamer of things. We could easily collect many texts concerning the subtle anxiety of gentle light. The flame of the candle is oracular. Let me give one brief example.

During a night of terror, Strindberg's lamp is smoking:

> I got up and opened the window. A draught of cold air threatened to put out the lamp. I closed it again.
> The lamp began to sing, to groan, to whimper.[7]

Let us remember that Strindberg originally wrote this story in French. Since the flame whines it has a child's grief, and so the entire universe is unhappy. Strindberg knows, once again, that all the beings of the world presage unhappiness for him. To whine, is this not to flicker in a minor key, with tear-filled eyes? And if the tears are heard in the voice, is not such a word onomatopoeic for the liquid flame that now and then is posited in the philosophy of fire?

In another passage from the same story,[8] Strindberg suspects an ill will on the part of the light: it is a noise from the candle that presages unhappiness:[9]

> I lit candles in order to pass the time reading. A sinister silence held sway and I could hear the beating of my heart. Then—a tiny noise that sounded like an electric spark.
> Could it be?

---

7. August Strindberg, *Inferno*, trans. Derek Coltman and Evert Sprinchorn in *Inferno, Alone and Other Writings* (Garden City, 1968), 246.
8. Ibid., 258.
9. "In Lombardy, the crackling of the embers and the creaking of the log are fatal portents." Angelo de Gubernatis, *Mythologie des plantes; ou, Les légendes du règne végétal*, I: 266.

> An enormous lump of tallow had fallen from one of the candles onto the floor. Nothing more than that. But where I came from it was an omen of death.

Certainly Strindberg is a thin-skinned soul. He is sensitive to the tiniest dramas of matter. The coke in his hearth alarms him when it crumbles too much while burning, when its residues do not fuse. But the disaster is at once greater and more subtle when it comes from the light. Do not lamp and candle give off the most humanized fire? In giving light, is not fire the creator of the greatest values? A disturbance at the peak of natural values lacerates the heart of a dreamer who would be at peace with the universe.

Note that no trace of symbolism is evident in Strindberg's anxiety over the candle's misery. The incident is everything. Though small, it shows itself to be a relief map of reality.

It is easy to show how childish this vesania is, yet it is still surprising that it occurs in a story full of real domestic suffering. But the fact is there; the psychological fact lived by the writer is doubled by the literary fact. Strindberg is confident that an insignificant event is capable of shaking the human heart. He thinks that with a slight scare he can inject fear into the reader's solitude.

Of course a psychiatrist has no trouble diagnosing schizophrenia when reading Strindberg's stories. But in their literary form such stories pose a problem: are they not schizogenic? When reading *Inferno* attentively, will not every reader have his moments of schizophrenia? Strindberg knows that he communicates with the great Other of solitary readers by writing in the absolute of solitude. He knows that in every soul there is an area where, beyond reason, the most childish fears survive. He will surely be able to promulgate his candle miseries. In *Inferno*, he follows the axiom expressed in his autobiography, "Go there and the others will fear."[10]

---

10. Strindberg, *L'Ecrivain*, trans. Stock, 167.

V

When a fly plunges into the flame of a candle the sacrifice is noisy—wings crackle, the flame leaps up. It seems that life crackles in the heart of the dreamer.

The death of the moth is smoother, not so noisy. It flies soundlessly; it barely touches the flame and already it is consumed. For a dreamer who dreams grandly, the simpler the incident the farther the commentaries go. To exhibit this drama C. G. Jung wrote an entire chapter entitled "The Song of the Moth."[11] He cites a poem by Miss Miller, a schizophrenic whose case provides the point of departure for the first edition of the *Metamorphoses of the Soul*.

Here again poetry gives a sense of destiny to an insignificant event. The poem enlarges everything. Coiled for a long time in its chrysalis, it is toward the sun—the flame of flames—that this tiny creature goes in search of the supreme and glorious sacrifice.

Here is how the moth, the schizophrenic, sings:

> I longed for thee when first I crawled to consciousness.
> My dreams were all of thee when in the chrysalis I lay.
> Oft myriads of my kind beat out their lives
> Against some feeble spark once caught from thee.
> And one hour more, and my poor life is gone;
> Yet my last effort, as my first desire, shall be
> But to approach thy glory; then, having gained
> One raptured glance, I'll die content,
> For I, the source of beauty, warmth, and life
> Have in its perfect splendor once beheld!

Such is the song of the moth, the symbol of a dreamer who wants to die in the sun. And Jung does not hesitate to associate his schizophrenic's poem with Faust's dream of losing himself in the light of the sun:

---

11. Jung, *Symbols of Transformation*, vol. 5 in *Collected Works of C.G.Jung*, trans. R.F.C. Hull (Princeton, 1956), 79.

> On tireless wings uplifted from the ground.
> Then should I see, in deathless evening light,
> The world in cradled stillness at my feet. . . .
> Yet stirs my heart with new-awakened might,
> The streams of quenchless light I long to drink.[12]

We do not hesitate to follow Jung as he associates the schizophrenic's poem with Goethe's because there we experience that *amplification* of image which is one of the most constant dynamics of literary reverie. For us it is a witness to the psychological dignity of the written reverie.

In *The Divan*,[13] Goethe takes as his theme *selige Sehnsucht*, blessed nostalgia, the sacrifice of the butterfly in the flame:

> I praise what is truly alive,
> what longs to be burned to death.
> In the calm water of love-nights,
>
> . . . .
>
> a strange feeling comes over you
> when you see the silent candle burning.
> Now you are no longer caught
> in the obsession with darkness,
> and a desire for higher love-making
> sweeps you upward.
>
> . . . .
>
> now, arriving in magic, flying,
> and, finally, insane for the light,
> you are the butterfly and you are gone.

Goethe provided a great motto for such a destiny: "Die and become."

> And so long as you haven't experienced
> this: to die and so to grow,

---

12. Ibid., 80.
13. Johann Wolfgang von Goethe, "The Holy Longing," trans. Robert Bly in *News of the Universe: Poems of Twofold Consciousness* (San Francisco: 1980), 70.

> you are only a troubled guest
> on the dark earth.

In his preface to *The Divan*,[14] Henri Lichtenberger provides a broad commentary on the poem. The mysticism of Oriental poetry

> appeared to Goethe to be related to ancient mysticism, to Platonic and Heraclitean philosophy. Goethe, who plunged into the reading of Plato and Plotinus, distinctly perceived the relationship between Greek symbolism and Oriental symbolism. He recognized the identity between the Sufi theme of the butterfly throwing itself into the flame of the torch and the Greek myth which makes the butterfly the symbol of the soul, showing us Psyche in the form of a young girl or a butterfly, seized and captured by Eros, burned by the torch.

VI

The moth plunges into the flame of the candle: "positive phototropism," says the psychologist who measures material forces; "Empedocles complex," says the psychiatrist who wants to find something human at the root of primary impulses. And both are right. But it is reverie that makes them agree, since the dreamer, watching the moth in his tropism, his death instinct, faces the image and says, why not me? Since the moth is a tiny Empedocles, why will I not be a Faustian Empedocles who, in death by fire, in the sun, will conquer light?

Though the butterfly may come to burn its wings in the lamp because we did not take the trouble, before the incident could occur, of extinguishing the flame, this cosmic lapse rouses no sensitivity in us. Still, what a symbol this is, this being who burns up his wings! Burning his raiment, burning his being—a dreaming soul will never stop meditating upon it. When Pierre-Jean Jouve's Paulina sees her own beauty

---

14. Goethe, *Le Divan*, trans. Lichtenberger, 45-46.

CHAPTER TWO • 35

before her first ball, when she wants to be as pure as a nun and yet tempt all the men, it is the death of a butterfly in the flame that she evokes:

> No, no, dear butterfly, watch out for the flame, another one is going to die like the one last night, it's going to die right now! It's coming back to the fire in spite of itself, it doesn't understand fire, half of one wing already burned, it comes back, and comes back again, but that's fire, poor butterfly, that's fire.[15]

Paulina is a pure flame, but a flame for all that. She wants to be a temptation, but she herself is tempted. She is so beautiful! Her own beauty is a fire that tempts her. From the first scene the drama of the death of purity in sin is played out. Jouve's novel is the story of a destiny. To die by love, in love, like the butterfly in the flame, isn't this to realize the synthesis of Eros and Thanatos? Jouve's account is animated simultaneously by life and death instincts. These two instincts, as revealed by Jouve in their depth and primitiveness, are not opposites. Jouve as depth psychologist shows that they operate in the rhythms of one particular destiny, in those rhythms that cause endless revolutions in a single life.

And the primal image of feminine destiny which Jouve chooses is that of a butterfly burned by a candle on the night of the first ball.

I have wished to pay attention to the most diverse flame dreamers, even those who meditate on the death of Luna moths attracted by the light. But I do not join in such reveries. I have experienced vertigo. The void attracts me and frightens me. But I do not suffer from an Empedoclean vertigo.

The solitude of death is too great a subject of meditation for the dreamer of solitude that I am. To finish this chapter I must tell once again how the simple and tranquil reveries evoked at its beginning were made my own.

15. Pierre-Jean Jouve, *Paulina 1880*, trans. Rosette Letellier and Robert Bullen (Indianapolis, 1973), 24.

VII

Jean Cassou always dreamed of approaching the great poet Oscar Vladislas Milosz with a question worthy of royalty, "How fares Your Solitude?"

This question has a million answers. In what center of the soul, in what corner of the heart, in what turn of the mind, is a great solitary man alone, really alone? Alone? Shut away or barricaded? In what refuge, in what cell is the poet really a solitary? And when everything alters according to the disposition of the sky and the color of dreams, each impression of the great solitary man's solitude must find its own image. Such "impressions" are first of all images. One must imagine solitude to know it—to love it or defend oneself from it, to be tranquil or to be courageous. He who wishes to create a psychology of the psyche's chiaroscuro in which the consciousness of our being is illuminated or obscured must multiply images and duplicate each one. A solitary man, in the glory of being alone, often believes that he can say what solitude is. But to each his own solitude. And the dreamer of solitude can only give us a few pages in this scrapbook of the chiaroscuro of solitude.

As for me, in total sympathy with the images poets offer me and with others' solitude, I come to be alone by means of their solitude.

I become alone, profoundly alone, from the solitude of another.

But of course this invitation to solitude must be discreet; it must be precisely a solitude of imagery. If the solitary writer wants to tell me his life, his whole life, he quickly becomes a stranger to me. The reasons for his solitude will never be the reasons for my solitude. Solitude has no history. All my solitude is contained in a primary image.

Here, then, is the simple image, the central picture in the chiaroscuro of dreams and memory. The dreamer is at his table, in his garret. He lights his lamp, a taper, his candle. Then I remember; then I find myself again. I keep the same vigil as he; I study as he studies. The world is for me as for him a difficult book illuminated by the flame of a candle. For the candle, companion of solitude, is above all the companion of solitary work. The candle does not illuminate an empty room; it illuminates a book.

Alone at night, with a book illuminated by a candle—book and candle, double island of light against the double darkness of the night and of the mind.

I study! I am only the subject of the verb "to study."

I dare not think.

Before thinking, one must study.

Only philosophers think before studying.

But the candle will burn out before the difficult book is understood. One should not waste the time of candlelight, the long hours of the studious life.

If I raise my eyes from the book to look at the candle, I dream instead of studying.

So the hours ripple in the solitary vigil, undulate between the responsibility of knowledge and the freedom of reveries, that too easy freedom of a solitary man.

I myself need no more than the image of a person at his candlelight vigil to begin this undulant movement of thoughts and reveries. Indeed I would be troubled if the dreamer at the center of the image told me the reasons for his solitude, some distant history of life's betrayals. Ah, my own past is enough to burden me. I have no need for someone else's. But I need others' images to revivify my own. I need others' reveries to remind me of my work in the faint light, to remind me that I too have been a candle dreamer.

3

The Verticality of Flames

> Above . . . light drops its clothes.
> OCTAVIO PAZ, *Eagle or Sun?*

I

AMONG THE REVERIES that make us feel lighter, reveries of height are the simplest and most effective. All upright objects point to a zenith. An upright form soars up and carries us along in its verticality. Conquering a real summit remains on the level of an athletic feat. The dream goes higher; it carries us into the beyond, into verticality. Many dreams of flight are born from an attempt to rival the verticality possessed by upright and vertical things. Near towers and trees a dreamer of heights dreams of the sky. Reveries of height nourish our instinct for verticality, an instinct repressed by the obligations of common life, of flatly horizontal life. Verticalizing reverie is the most liberating of reveries. There is no surer way of dreaming well than to dream of somewhere else. But of all such places is not this place *somewhere above us* the critical one? Dreams occur in which "above" forgets or suppresses "below." Living at the zenith of the upright object, gathering reveries of verticality, we experience a transcendence of being. Images of verticality make us enter the realm of values. To communicate with the verticality of an upright object through the imagination is to enjoy the advantages of ascensional forces. It is to participate in the hidden fire that dwells in beautiful forms, forms self-assured in their verticality.

I developed this theme of verticality extensively in a chapter of *Air and Dreams*.¹ In this chapter can be seen the entire background of our present reveries on the flame's verticality.

II

The simpler their subject the greater the reveries. The flame of the candle on the table of a solitary man prepares the way for all reveries of verticality. The flame is a valiant and fragile verticality. A breath disturbs it, but it rights itself. Ascensional force reasserts its own prestige.

A line from Trakl says,

> The candle burns high and its crimson rears up.²

The flame is an inhabited verticality. Everyone who dreams of a flame knows that the flame is alive. Its verticality is guaranteed by its sensitive reflexes. As soon as an accident of combustion disturbs the élan of its zenith, the flame reacts. One who dreams of the verticalizing will, who learns the lesson of the flame, realizes that he must right himself. He rediscovers the will to burn high, to go with all his strength to the summit of fervor.

And what a great moment, what a beautiful moment results when the candle burns properly! What delicacy of life in the flame that lengthens and tapers off! Then the values of life and of the dream are united.

> A stem of fire. Are we ever aware of all that is fragrant?

says the poet.³

Yes, the stem of the flame is so straight and so frail that the flame is a flower.

Images and things thus exchange their properties. For

---

1. *Air and Dreams* (Dallas, 1988), chapters 1 and 4.
2. *Anthologie de la poésie allemand*, trans. Stock, II:109.
3. Edmond Jabès, *Les Mots tracent*, 15.

someone dreaming of a flame, the entire room takes on an atmosphere of verticality. A gentle but unfailing dynamism draws dreams to their summit. One may well be interested in the innermost vortex that surrounds the wick, see in the body of the flame the stirrings of the struggle between darkness and light. But everyone who dreams of a flame elevates his dream toward the summit. There fire becomes light. Villiers de l'Isle-Adam used this Arabian proverb as epigraph for a chapter of his *Isis*: "The torch does not illuminate its base."

The greatest dreams are at the summit.

The flame is so essentially vertical that it appears, for one who dreams about existence, to be stretched toward transcendence, toward an ethereal non-existence. In a poem entitled "Flame" we read:[4]

> Bridge of fire flung between real and unreal
> continuous co-existence of being and non-being.

To play at being and non-being with a mere nothing, with a flame, and perhaps only an imaginary one at that, is, for a philosopher, a fine moment of illustrated metaphysics.

But every soul with depth has its personal realm of transcendence. The flame illustrates every form of transcendence. Looking at a flame Claudel wonders: "Whence matter takes flight to wing itself into the category of the divine."[5]

If I had accorded myself the right to meditate on liturgical themes, I would have no trouble finding documents concerning the symbolism of flames. I would then have to face scholarship. I would go beyond the plan of my little book, which must be content to grasp symbols in their outlines. Whoever wants to enter the world of symbols that fall under the sign of

---

4. Roger Asselineau, *Poésies incomplètes*, Éd. Debresse, 38.
5. Paul Claudel, *The Eye Listens*, trans. Elsie Pell (New York, 1950), 154.

fire can refer to the great work of Carl-Martin Edsman: *Ignis divinus*.[6]

III

In my preliminary chapter I refused to worry over scholarship and scientific or pseudo-scientific experiments with the phenomena of flames. I did my best to stay within the homogeneity of reveries that imagine, reveries of a solitary dreamer. One cannot be two persons and dream profoundly of a flame. The artless observations Goethe and Eckermann made together as master and disciple do not lead to thought; their observations cannot be repeated with the seriousness appropriate to scientific research. Nor do they provide us an opening onto that philosophy of the cosmos which had such a great effect upon German romanticism.[7]

To show immediately how with Novalis one moves from physics based on facts to physics based on values, I shall offer a commentary on the short motto that appears in the Minor edition:[8] "*Licht macht Feuer*," light makes fire. In its German form this three-syllable sentence goes so quickly, it is such a rapid arrow of thought, that common sense does not immediately feel its wound. All we know from everyday life enjoins us to read the sentence in reverse; normally one lights a fire in order to make light. The challenge of the sentence can be met only by adhering to a cosmology of values. The three-syllable sentence "*Licht macht Feuer*" is the first act of an idealistic revolution in the phenomenology of the flame. It is one of the pivotal sentences that a dreamer repeats to himself to condense his conviction. I imagine I hear for hours on end the three syllables on the poet's lips.

---

6. Carl Martin Edsman, *Ignis divinus; le feu comme moyen de rajeunissement le d'immortalité: contes, légendes, mythes et rites* (Lund, 1949). By the same author: *Le Baptême du feu* (Uppsala, 1940).
7. Cf. *Conversations de Goethe et d'Eckermann*, I:203, 255, 258, 259.
8. III:33.

An idealist proof would not be out of the way: for Novalis the ideality of the light must explain the material action of fire.

Novalis' fragment continues: "*Licht ist der Genius des Feuerprozesses*"—"light is the genius of the fire process." This is the most serious of all statements for a poetics of material elements, since the primacy of light deprives fire of its power as absolute subject. Fire receives its real existence only at the conclusion of the process of becoming light, when, through the agonies of the flame, it has been freed of all its materiality.[9]

If this reversal of causality were deciphered in the flame, the tip would be the storehouse of its activity. Purified at the tip, light draws from the entire taper. The light, then, is the true driving force determining the ascensional being of the flame. The main principle of Novalis' idealizing cosmology is to take account of values in the activity itself; where they are more than facts they discover their being by rising. All idealists find, in meditating upon a flame, the same ascensional stimulation. Claude de Saint-Martin writes: "The movement of the mind is like that of fire. It is created as its rises."[10]

IV

Coordinating all the fragments in which Novalis evokes the verticality of flames, one could say that everything upright, everything vertical in the cosmos, is a flame. Speaking more dynamically we should say: everything that rises possesses a flame's dynamics. The reverse, only slightly less exaggerated, is clear. Novalis writes: "In the flame of a lamp all natural forces are active." (*In der Flamme eines Lichtes sind alle Naturkräften tätig*).[11]

Flames constitute the very being of animal life. And in-

---

9. For one author of the Encyclopedia (article: Feu, 184): "A bright, lively flame (gives more heat) than the hottest furnace."
10. Claude de Saint-Martin, *Le Nouvel homme*, an IV, 28.
11. Novalis, *The Novices of Sais*, trans. R. Manheim (New York, 1949), 95.

versely Novalis notes "the animal nature of the flame."[12] The flame is in some way a naked animality, a kind of excessive animal. It is the glutton par excellence (*das Gefrässige*). These aphorisms, fragments found everywhere in his work, demonstrate the direct character of his convictions. These are reverie truths that only an experience of profound oneirism, resulting more from dreaming than reflecting, can demonstrate.

Each of the biological kingdoms of life consequently corresponds to a particular type of flame. In the fragments translated by Maeterlinck we read: "The tree can only become a flowering flame, man a speaking flame, the animal a wandering flame."[13]

Paul Claudel, apparently without having read Novalis' text, writes similar lines. For him life is a fire. Life stores its fuel in the plant, which is ignited in the animal: "The vegetable, or elaboration of the combustible matter. The animal or the condition of ignited matter providing for its own living," says Claudel in the summary that is prelude to his story.[14] "If the vegetable can be defined as 'the combustible substance,' the animal is burning substance."[15] "The animal maintains its form by burning whatever it needs to satisfy the hunger for fire concealed in it."[16]

For Novalis as well as Claudel the dogmatic tone of this motto-like cosmology will keep the erudite philosopher at a

---

12. Éd. Minor, II:206.
13. Cf. a single passage in which everything that lives is given as the excrement of a flame. We are but the residues of an inflamed being (Ed. Minor, II:216).
    In *West-eastern Divan*, Goethe writes:
    > There by the hearth, where fire has virtuous use,
    > Mellow of beast and plant the sap or juice. (171)
    > An des Herdes raschen Feuerkräften
    > Reift das Rohe Tier- und Pflanzensäften
14. Paul Claudel, *Poetic Art*, trans. R. Spodheim (New York, 1948), 55.
15. Ibid., 59.
16. Ibid., 60.

distance. It is different when such aphorisms are set in the framework of a poetics. Here the flame is creative. It gives us poetic intuitions that make us participate in the inflamed life of the world. The flame is thus a living substance, a poeticizing substance.

The most diverse beings are made substantive by the flame. Only an adjective is necessary to make them more specific. A cursory reader will perhaps see no more here than stylistic play. But if he participates in the inflammatory intuition of a poetic philosopher, he will understand that the flame is the source for a living creature. Life is a fire. To know its essence one must burn in communion with the poet. To use an expression of Henry Corbin, we could say that Novalis' formulas tend to raise meditation to incandescence.

V

But there is a dynamic image through which a meditation on flames gains a sort of *super-élan vital* that can heighten life, prolong it beyond itself, despite all the failings of ordinary matter. Novalis' fragment 271 summarizes an entire philosophy of flame-life, of the life-flame:[17]

> The art of leaping beyond oneself is everywhere the highest act. It is life's point of origin, the genesis of life. The flame is nothing other than an action of this sort. Thus philosophy begins at the point where the philosophizer philosophizes himself, that is to say, where he consumes and renews himself.[18]

In a revised version of his text Novalis brings together the two meanings of the verb *verzehren* (to consume and to consummate). He points to the transition in flame activity, from

---

17. Novalis, Éd. Minor, II:259.
18. Cf.Nietzsche, "Bruchstucke zu Dionysos-Dithyramben" in *Werke*, band 2 (Leipzig, 1930), 578:
> Life itself has created for itself
> Its supreme obstacle.
> Now it leaps over its own thought.

determined to determining, from a satisfied being to one that lives out its freedom. A being becomes free by consuming itself in order to be renewed, by taking on the destiny of a flame, and especially by welcoming the destiny of a super-flame which glistens above its very tip.

But before philosophizing perhaps one should look again; perhaps, failing that, one should re-imagine that rare phenomenon of the fireplace—the peaceful flame which throws off flying sparks that float lighter and freer within the cloak of the chimney.

I have often seen this spectacle in dreamy vigils. Sometimes my good grandmother, with a hemp stalk deftly held above the flame, would rekindle the slow smoke that ascended the length of the blackened hearth. The lazy fire does not always burn all the elixirs of the wood at once. Smoke leaves the shining flame reluctantly; the flame had much more to burn. In life too there are many things to re-inflame!

And when the super-flame had come back to life, my grandmother would say to me: "See, my child, these are the fire birds." So I, further elaborating in dreams what she had said, believed that these birds of fire had their nest at the heart of the log, well hidden beneath the bark and the tender wood. This nesting place—the tree in the course of its growth—had created this cozy nest where the beautiful fire birds nestled. In the heat of a great fireplace, time had just hatched and flown away.

I would be dubious about telling my own dreams and distant memories if the primary image, the flame that leaps above itself and continues burning, were not a true image. Charles Nodier saw this flame that overleaps itself, that has a new momentum beyond its first momentum, beyond its tip. He speaks of "those dreamed-of fires that fly above torches and candelabra, when the cinder that produced them has already cooled."[19]

---

19. Charles Nodier, *Oeuvres complètes*, V:5.

CHAPTER THREE • 47

For Nodier this surviving, hovering [*survivante, survolante*] flame illustrates a remote comparison. He speaks of a time when "only love lived above the social world, like those fires that make a purer light above the torches."

For a Novalis dreamer of animalized flames, the flame, because it flies, is a bird.

> Where else will you take the bird
> Than into the flame?

asks a young poet.[20]

In my dreams and in my games at the hearth, I was thus quite familiar with this domestic Phoenix, the most ethereal Phoenix of all, not reborn from his ashes but from his mere smoke.

But when a rare phenomenon is the basis of an extraordinary image, an image that fills the soul with immeasurable dreams, to whom, or to what, must it provide reality?

A physicist will answer: Faraday used this experiment with a candle relit from its own smoke as the subject of a popular lecture.[21] This lecture is one among several that Faraday gave during some evening courses and collected under the title *History of a Candle*. To perform the experiment one must blow very softly on the candle, then quickly relight the vapor—and only the vapor—without disturbing the wick.

Half-knowing, half-dreaming, I would answer in this way: to perform Faraday's experiment successfully one must act quickly, because real things do not dream long. One must not allow the light to fall asleep. One must hasten to awaken it.

---

20. Pierre Garnier, *Roger Toulouse*, Cahiers de Rochefort, 40.
21. Michael Faraday, *The Chemical History of a Candle* in *Scientific Papers*, The Harvard Classics, ed. Charles W. Eliot (New York, 1910), 89-180.

4

Poetic Images of the Flame in Plant Life

> I no longer know if I am sleeping
> For the light is vigilant in the heliotrope.
> CÉLINE ARNOLD, *Anthologie*

I

WHEN DREAMING a little of the forces that maintain each object's particular form, one can easily imagine that in each vertical being there reigns a flame. That is, the flame is the dynamic element of upright life. I previously cited a line from Novalis: "The tree is nothing other than a flowering flame," and this theme I will now illustrate by recalling the images that are endlessly reborn in the imaginations of poets.

Before recounting the exploits of the poetic imagination, perhaps it is necessary to repeat that a *comparison* is not an *image*. When Blaise de Vigenère *compares* the tree to a flame, he simply brings words together without really succeeding in harmonizing plant and flame vocabulary. Let us cite a passage that appears to be a good example of a prolix comparison.

Hardly has Vigenère spoken of the flame of a candle than he speaks of the tree:

> In a similar way [to the flame], it has its roots fixed in the earth, from which it takes its nourishment as the candle-end does from the tallow, wax, or oil that makes it burn. The trunk that draws its juice or sap is like the candle-end, where the fire subsists on the liquid that it attracts to itself. The white flame is the branches and boughs adorned with leaves; the flowers and fruits, which manifest the final end of the tree, are the white flame to which everything is reduced.[1]

---

1. Vigenère, 17.

Throughout this extended comparison, we never grasp a single one of the million igneous secrets that remotely account for the flamboyant explosion of a flowering tree.

Guided by poets, I shall try to seize on images in primal poetry born of details worthy of elaboration, seeds of living poetry that we can make live within us.

II

When the image of the flame is thrust on a poet with its truth about the plant world, it must consist of a single phrase. To explicate it, to develop it, would be to drag it out, to halt the momentum of an imagination unifying the intensity of fire and the patient strength of foliage. Image-phrases that tell of plant flames, that paint them, are just so many polemical acts for countering common sense, with its dull ways of seeing and speaking. But with a new image the imagination is so certain of possessing one of the world's truths that a polemic against the unimaginative would be a waste of time. It would be better for the imaginer speaking to imaginers to make fresh statements again and again about the flames of plant life.

Thus begins the reign of decisive images, poetic decisions. All poetry is a beginning. I propose to designate these image-phrases, pregnant with the will for new expression, by the name *poetic maxims*. The term "fragments," used by fragmenters, is unjust. Nothing is fragmented in an image that finds strength through condensation.

With a dictionary of beautiful maxims from the dogmatic imagination, a botany of all the plant-flames cultivated by poets, one might perhaps decipher the dialogue of the poet with the world. It will always be difficult to give order to a large number of deliberately unusual images. But sometimes an alluring passage, some odd image, is enough to show the relationship between two different genres. For example, how can one not have the impression that Victor Hugo and Balzac

CHAPTER FOUR • 51

belong to the same family of dream botanists when these two poetic sentences are placed side by side?

> Every plant is a lamp. Its perfume is its light.[2]
> Every smell is a combination of air and light.[3]

Of course in Balzac's aesthetics, it is the plant at its peak—the flower—that realizes this prodigious synthesis of air and light.

A sort of Baudelairean correspondence is activated by heights and summits, as though what is of value at the summit stimulated what is of value at the base. Thus the dreamers who live out in both directions this correspondence between fragrance and light read with conviction a "thought" that gives value to delicate light: "Certain trees become more fragrant when they are touched by the rainbow."[4]

III

From an outstanding poet one can receive the very germ of an image that is even more concentrated than a poetic maxim, an image-seed or a seed-image. The following bears witness to a flame that burns in the innermost being of the tree, where all the promise of flamboyant life resides. Louis Guillaume, in a poem entitled "The Old Oak,"[5] with three words fills us with reveries: "Pyre of sap," he says, in order to glorify the great tree.

"Pyre of sap"—words never before spoken, the sacred seed of a new language that must envisage the world poetically. The poetic maxim is left to the reader's care. A million poetic maxims will be dreamed of in this igneous sap that gives the strength of fire to the king of trees. As for me, awakened from my old images by the gift of the poet, I leave the great image

2. Victor Hugo, *The Man Who Laughs* (Boston, n.d.), II: 37.
3. *Louis Lambert* in *The Works of Honoré de Balzac* (New York, n.d.), II:101.
4. Le sieur de La Chambre, *Iris*, 20.
5. Louis Guillaume, *La Nuit parle*, éd. Subervie, 28.

of the great being twisted in sufferings like Laocoön, and, dreaming of this sap that rises and burns, I sense that the tree is a fire-bearer. And a great destiny for the oak is predicted by the poet. This oak is the vegetable Heracles who, in every fiber of his being, prepares for his apotheosis in the flames of a pyre.

A world of cosmic contradictions is born from this nexus of hostile powers. Louis Guillaume has linked fire and water together in three words. That is surely a triumph of language. Only poetic language can be so daring. Here we are truly in the domain of a free and creative imagination.

IV

Sometimes it is as if the image-seed is exuberant. It shoots up in one burst to the extremes of its splendor. In a single image Jean Caubère confers a flame-like meaning on a solitary water jet, that upright being more upright than all the trees of the garden. "Caubère's *water jet*"—what a great privilege it is to give one's name to an uncreated image—it is, for me, the flame of spirited water, the fire that splashes to its full height, to the limit of its upright action.[6]

> There are gardens
> where burns a solitary water jet
> among the rocks
> at dusk.

The poet provides us with the enormous pleasure of speech. Through him we transcend elementary differences. The water burns. It is cold but it is strong, and so it burns. It gains the virtue of an imaginary fire by a sort of natural surrealism. Nothing is deliberate; nothing is fabricated by this *immediate surrealism* of the water jet-flame. Jean Caubère has concentrated the surrealism of his image into a single word: the word *burn*, which de-realizes and surrealizes. And this single word

6. Jean Caubère, *Déserts, poèmes.* Éd. Debresse, 18.

burn has overthrown the poem's crepuscular melancholy. The image achieved is thus a witness of creative melancholy.

Such a synthesis, such a fusion of objects embodied in such different forms—like the synthesis of water jet and flame, of tree and flame—could hardly have been expressed in prose. It needs the poem and its suppleness, its poetic transmutations. A hymn seizes upon the being of images and makes of them a subject, a hymnal subject. It is the hymn itself that is the synthesizing power. The Mexican poet Octavio Paz knows this well, saying very precisely that the hymn is at once

> Poplar of fire, water jet.[7]

Here again the poet leaves to the reader the task of interpolating phrases—that poetic pleasure of writing poetic maxims which unite the flame of the tree launched upwards and the vertical flame of the water jet. With the poets of our times, we enter into a realm of blunt poetry that does not chatter away but always desires to live in primal speech. We must listen to poems as though they were words heard for the first time. Poetry is wonder, quite precisely at the level of speech, in speech, and by speech.

I shall seize on every opportunity to speak enthusiastically of autonomous poetic values. But I must return to my more specific intentions of studying the plant images of the flame by arriving at simpler examples of the relationship between lights, flowers, and fruits.

V

> A tree is much more than a tree

says a poet.[8]

---

7. Octavio Paz, *Aigle ou Soleil?*, 83.
8. Gilbert Socard, *Fidèle au monde, poèmes*, 18.

It rises toward the most precious light of its being, and that is why in many poems fruit-bearing trees are lamp-bearers. The image is very natural in the poetry of gardens. All of these lights in the summer foliage nourish fire. One of Dickens' characters confides that when he was a child he thought "that birds' eyes were so bright, perhaps, because the berries that they lived on in the winter were so bright."[9]

In a lecture on Matisse's paintings entitled *The Poetry of Light*, Arsène Soreil quotes an Oriental poet who said:

> Oranges are the lamps of the garden.

Soreil also quotes Marcel Thiry:

> In the apple trees the fruits shine like lamps.

But these images are too brief; they are terminal; they do not follow the course of long reveries that see in the transforming tree the sap of life as a substance of fire and flame.

When the August sun has fermented the first sap, fire slowly appears in the grape cluster. The grape brightens. The cluster becomes a brightness that shines under the shade of the large leaves. The modest vine leaf must have found its first use in hiding the cluster.

Ascension of fire, ascension of light—the poets of cosmic reveries choose between these two images. For Rachilde in his youth, the vine, absorbing all the fires of the earth through its virile stock, gives the cluster "this satanic sugar distilled in the violence of the volcano."[10]

Man's intoxication brings the madness of the vine to its fulfillment.

---

9. Dickens, "The Haunted Man and the Ghost's Bargain" in *Christmas Stories* (Philadelphia, 1843), 132.

10. Rachilde, *Contes et nouvelles, suivis du théâtre* (Paris: Le Mercure de France, 1900), 150.

A poet tells of the union of three movements in every tree:

> Tree fountain, tree spout, arch of fire.[11]

There are trees that have fire in their buds. For d'Annunzio the laurel is a tree so warm that when pruned its trunk is soon covered with buds like "green sparks."[12]

VI

A Novalis dreamer will easily accept the following formula as one of the axioms of the vegetable world's poetics: flowers, all flowers, are flames—flames that want to be light.

This becoming light is what every dreamer of flowers feels; it is brought to life by going beyond what one sees, going beyond reality. The poet-dreamer lives in the radiance of all beauty, the reality of unreality. The poet who does not have the painter's privilege of creating with colors has no interest in rivaling the glamour of painting. Caught by the rigor of his discipline, the poet, this painter in words, knows the glamour of freedom. He must speak the flower, say the flower. He can only understand the flower by enkindling the flames of the flower with the flames of speech. Poetic expression is, then, that evolution of light which every Novalis dreamer has sensed in his philosophic contemplations.

The poet's problem is thus to express the real by means of the unreal. As I indicated in my preface, he lives in the chiaroscuro of his being, by turns bringing a glimmer of light or a penumbra to the real—and each time giving an unexpected nuance to its expression.

But let us "look into" some poetic expressions of various flower-flames subtly colored according to the poet's genius.

---

11. ¿Aguila o sol?/Eagle or Sun, trans. Eliot Weinberger (New York, 1976), 113.
12. D'Annunzio, La Contemplation de la mort, trans. Doderet, Calmann-Lévy, 59.

We may begin with images in which the flower-flames might have been borrowed, like reflections of a setting sun:

> The sky grows dim and the chestnut trees burn,

writes Jean Bourdeillette.[13]

The tall foliage of autumn chestnut trees plays its part in the symphony of the setting sun. Thus, if the poem is taken in its totality, it is easy to imagine that every tree has its own effect of light. The blazing fire at the heights comes down into every flower of the garden. Bourdeillette's poem ends with this grand line:

> The dahlias have kept the embers of the sun.

When I read such a poem pyrophorically, I can feel the *unity of fire* between sun, tree, and flower.

Unity of fire? Even the unity of action is conferred upon the world by poetic expression.

In the work of the same poet there are other more individualized flame-flowers. Is not a red tulip a cup of fire? Is not every flower a type of flame?

> Tulips of copper
> Tulips of fire
> Twisted in the warmth
> Of this month of May.[14]

If you put the garden tulip on your table, you have a lamp. Put a red tulip, just one, in a long-necked vase. Sitting by it in the solitude of a solitary flower, you will have candle reveries.

In a note Bernardin de Saint-Pierre writes: "Chardin says that in Persia when a young man presents a tulip to his mis-

13. Jean Bourdeillette, *Les Étoiles dans la main*, Éd. Seghers, 1954, 21.
14. Bourdeillette, *Reliques des songes*, éd. Seghers, 48.

tress, he wants her to understand that he, like this flower, has a countenance of fire and a heart of glowing coal."¹⁵ Indeed, at the bottom of the calyx the wick of the torch is completely blackened.

When the flower is a quiet lamp, an undramatic flame, the poet finds words which are the bliss of speech:

> The blue lupins burned
> Like soft lamps.¹⁶

Here, in the realm of speech itself, is a moist flame that flows in its labial syllables.

I imagine a beautiful, tender woman who repeats these two lines again and again while looking at herself in her mirror. Her lips would be happy. Her lips would learn to flower gently.

Of all flowers, the rose is a veritable image-hearth for imaginary plant flames. It is the very embodiment of imagination eager to be convinced. What intensity there is in this single line by a poet who dreams of a time when

> . . . the fire and the rose are one.¹⁷

In order to give double value to each image they must harmonize in both directions. A dreamer of roses must see an entire rosebush in his fireplace.

Sometimes flowers seem to be born in the flaming coals. Thus Pieyre de Mandiargues writes:

> The fire of geraniums illuminates the coal.¹⁸

---

15. Jacques Henri Bernardin de Saint-Pierre, *Etudes de la nature* (Paris, 1791), II: 373.
16. Bourdeillete, *Reliques des Songes*, 34.
17. T. S. Eliot, *Four Quartets* in *The Complete Poems and Plays 1909-1950* (New York, 1971), 145.
18. *Les incongruités monumentales*, Éd. R. Laffont, 33.

What is the origin of this great red and black dream—the flower or the fireplace? For me, the poet's image works both ways, and in both instances violently.

Everything depends on the poet's temperament. For Lundkvist it is the placid cornflower: "the cornflower rises, electrically, in the cornfield and threatens the harvester like the flame of a blow-torch."

The lamp and the rose exchange their gentleness. Rodenbach, a man of gentle images, writes:

> The lamp in the bedroom is a white rose.[19]

In his house of a hundred mirrors Rodenbach cultivated imaginary flowers. Again he writes:

> The lamp
> that makes water lilies blossom in mirrors.

His reverie of reflection is so cosmogonic that he has created the vertical pool. In this way the poet covers the walls of his room with paintings of water lilies. Nothing can stop an imaginer who sees flowers in every light.

A more ardent poetic temperament will express the fire of roses with greater passion. D'Annunzio's work is rich in roses afire. One reads in the great novel *Fire*:

> . . . Look at the red roses.
> They are flaming. They seem to have a live coal at the heart. They are flaming really.[20]

This remark is so simple that it may seem trite to a reader skimming over it. But the writer wanted to present this as a dialogue of two lovers in the fire of passion. Red flowers can

---

19. Georges Rodenbach, *Le miroir du ciel natal*, 13.
20. D'Annunzio, *The Flame of Life*, trans. Kassandra Vivaria (Boston, 1900), 280.

put their stamp on someone's life. The dialogue resumes a few lines further:

> Look! they become more and more red. Bonifazio's velvet. . . .
> Do you remember? It is the same strength.[21]

The image reverses in another passage where d'Annunzio presents the work of glass blowers. It is molten glass that evokes the name of a flower—new proof of the reciprocal action between the two poles of a dual image: "the unformed vases trembled at the end of the irons, half rosy and bluish like clusters of hydrangea about to change colour."[22]

Correlatively, then, the fire flowers and the flower lights up.

These two corollaries could be developed endlessly: color is an epiphany of fire, and the flower is an ontophany of light.[23]

VII

When we experience the flower world, we are in a state of dissipated imagination. We hardly know how to receive them in all the intimacy of their being, as witnesses of a world of beauty, a world that multiplies its beautiful beings. Yet each flower has its own light. Every flower is a dawn. A dreamer of the sky should find in each flower a color of sky. So it would be in a reverie that through a will to life in the heights imparted super-Baudelairean correspondences to all things.

Henry Corbin, introducing a scholarly article entitled "Sympathy and Theopathy among the 'Fidèles d'amour' in Islam,"[24] quotes Proclus, as he alludes to "the heliotrope and its prayer":

> What other reason, asks Proclus, can be given for the fact that insofar as possible, the *heliotrope* follows the movement of the sun, and the *selenotrope* follows that of the moon like a funeral procession, like torches, unless we admit that there are causal

21. Ibid., 281.
22. Ibid., 301.
23. The first corollary is that of d'Annunzio.
24. In *Eranos Jahrbuch*, 1955, 199.

harmonies, intersecting causalities between beings of the earth and of the sky?

For truly everything prays according to the rank that it occupies in nature and sings the praise of the lord of the divine life to which it belongs—spiritual praise, and rational or physical or tangible praise; for the heliotrope moves according to its freedom of movement and in its course. If one were able to catch the sound of the air stirred by its movement, one would realize that it is a hymn to his king, such as a plant might sing it.

At what level, at what height should one meditate upon Proclus' text? Above all, it is necessary to feel that in order to achieve height Proclus develops all heights within himself. Fire, air, light, everything that rises also partakes of the divine; every dream unfolded is an integral part of the flower's being. The life flame of a being that flowers is a straining toward the world of pure light.

And all of these evolutions are the happy evolutions of slow movement. Torches in the gardens of the sky, in harmony with the flowers in men's gardens, are unfailing and slow flames. Sky and flowers harmonize to teach slow meditation, meditation that prays.

If we read further in Corbin's pages, we must open ourselves unreservedly to the dimension of Height—a Height that gains the dignity of the sacred. For Proclus, the heliotrope, in its heavenly color, prays because it is always turned toward its Lord in conspicuous fidelity. Henry Corbin cites this Koranic line: "Every being knows the manner of prayer and glorification that is right for him."[25] And Corbin shows that the tropism of the heliotrope is, for the Islamic "Fidèles d'amour," a *heliopathy*.

VIII

By dreaming naïvely of the images of the poets, we have accepted all the little miracles of the imagination. When

---

25. Ibid., 203.

CHAPTER FOUR • 61

poetic value is in play, it would be inappropriate to evoke other values and also inappropriate to approach our study with a mind that is the least bit critical. To finish this little chapter, however, let us look at a text that I cannot help but see through the eyes of a man from Champagne.

I take this anecdote from a very serious book. Sir James Frazer, without introduction or commentary, writes: "When the Menri came into contact with the Malays, they found among them a red flower (*gantogn*: Malay *gantang*). They gathered in a circle around it and stretched their arms over it to warm themselves."[26]

In what follows, the anecdote becomes more complicated. In particular, a deer and a green woodpecker come into play. The green woodpecker, epitome of legendary birds, can easily carry the fire within its brilliant feathers to the tribesmen. Frazer provides many texts concerning legendary animals that are the benefactors of man, and we learn to believe a little, just a little, in the legends that the ethnologists bring back to us. We become docile students in the school of naïveté. But in the story of this Malayan family assembled around a bouquet of ardent flowers to warm their fingers, the demon of irony seizes me and I reverse the axis of naïveté: how the eyes of the good savages must shine with mischief when they present to the missionary this comedy of the floral origin of fire!

26. Sir James George Frazer, *Myths of the Origin of Fire* (1930; rpt. ed.: New York, 1974), 100.

5

The Light of the Lamp

> Wishing to hearten a timid lamp
> great night lights all her stars.
> TAGORE, *Fireflies*
> (This short poem was written on
> a lady's fan.)

I

THE LIVED companionship of familiar objects leads us back to life that passes slowly. Close to them, we are caught up once more by a reverie that possesses its own past but is nevertheless fresh time and again. Objects kept on the whatnot shelf, in that tiny museum of beloved things, are the talismans of poetry. No sooner are they recalled than, through that grace which is their name, we are off and dreaming of an ancient story. What a disaster it is for reverie when the names, the old names of objects begin to change, to become attached to things completely different from the good old things on the old whatnot shelf! Those who lived in the century now past say the word *lamp* with lips different from those of today. Dreamer of words that I am, the word "lightbulb" makes me laugh. Never can a lightbulb be familiar enough to take a possessive adjective.[1] Who now can say "my electric lightbulb" in the same way that he once said "my lamp"? Ah! how are we to dream again, given the decline of possessive adjectives which once told so powerfully of the companionship we used to have with our objects?

---

1. Jean de Bosschère notes with quick sarcasm a scene in which, instead of the traditional oil-lamp, an "electric bulb" venerates the face of the Virgin Mary. Isn't the oil-lamp a look: "A small flame burned in the dark eye of its oil" (cf. *Marthe et l'engagé*, 221). The electric bulb has no look.

The electric lightbulb will never provoke in us the reveries of this living lamp which made light out of oil. We have entered an age of administered light. Our only role is to flip a switch. We are no more than the mechanical subject of a mechanical gesture. We cannot take advantage of this act to become, with legitimate pride, the subject of the verb "to light."

In his beautiful book *Toward a Cosmology*, Eugene Minkowski has written a chapter entitled "I Light the Lamp."[2] But the lamp here is an electric bulb. A finger on the switch is enough to make dark space immediately bright. The same mechanical gesture causes the inverse transformation. A little click says *yes* and *no* with the same voice. Thus the phenomenologist has the means of placing us alternately in two worlds, which is as much as to say "two consciousnesses." With an electric switch one can play the games of *yes* and *no* endlessly. But in accepting this mechanism, the phenomenologist has lost the phenomenological density of his act. Between the two universes of darkness and light, there is only one instant, a Bergsonian instant, an intellectual instant. That instant was more dramatic when the lamp was more human. When lighting the old lamp, one always feared that a mistake or mishap would occur. The wick is not quite the same this evening as it was yesterday. Without proper attention, it will blacken. If the glass is not straight, the lamp will smoke. One always gains something by giving familiar objects the attentive friendship they deserve.

II

It is in the friendship that poets have for things, their things, that we can know these burgeoning instants that give human value to ephemeral actions.

---

2. Eugene Minkowski, *Vers une cosmologie, fragments philosophiques*, éd. Aubier, 154.

Henri Bosco restores the lamp to its former dignity in those passages where he tells us his childhood memories. Of this lamp that remains true to our solitary being, he writes:

> One quickly perceives, not without emotion, *that it is someone*. By day, one believes it to be only a thing, something useful. But when the day fades and one wanders in a house alone, invaded by that penumbra which permits one to move about only by groping along walls, then the lamp that is sought, that cannot be found, that is discovered in the place where it had been forgotten; that lamp, seized, even before it has been lighted, reassures you and offers you a gentle presence. It calms you, it thinks of you . . . [3]

Such a passage will have few echoes for those phenomenologists who define the being of objects by their "implementarity." They have created this barbarous word to bring to an immediate stop the seductive power of things. "Implementarity" is for them such clear-cut knowledge that it has no need of the reverie of memories. But memories deepen our companionship with good objects, with faithful objects. Every evening, at the appointed hour, the lamp performs its "good deed" for us. These sentimental confusions between good object and good dreamer can easily draw criticism from the psychologist frozen in adulthood. For him they are only lingering aspects of childhood. But under the poet's pen, poetic meaning stirs again. The writer knows that his work will be read by souls sensitive to primal poetic realities. Bosco's passage continues:

> . . . Look well when you light it and tell me if, secretly, it doesn't light itself beneath our distracted eyes. Perhaps I would surprise you if I said that it does not so much receive the fire brought to it as offer us its flame. The fire comes from without. And this fire is only an occasion, a convenient pretext for the closed lamp to give off light. I can feel that it is a creature.

The word "creature" determines everything. The dreamer knows that this creature creates light. It is a creating creature.

---

3. Henri Bosco, *Un Oubli moins profond* (Paris: Gallimard, 1961), 316.

It is enough to give it its due, to remember that it is a good lamp and that it is alive. It lives in the memory of past peace. The dreamer remembers the good lamp that lit so well. The reflexive verb *s'allumait* [was lighted, lit itself] reinforces the value of the subject, the creature that gives light. Words, with their delicate flexibility, help us dream. Bestow value on things, from the bottom of your heart give active beings their rightful power, and the universe will be resplendent. A good lamp, a good wick, some good oil, and there you have a light that makes the human heart rejoice. Whoever loves a beautiful flame loves good oil. He follows the course of all cosmogonic reveries in which an object in the world is a seed of the world. For Novalis, oil is the very material of light; the beautiful yellow oil is condensed light, a condensed light that wants to spread out. Man has succeeded in liberating from this slight flame the forces of light imprisoned in matter.

Doubtless we no longer dream to this extent. But we once dreamed that way. We dreamed of the lamp which gives luminous life to dark matter. How could a dreamer of words not be moved when etymology teaches him that petroleum is petrified oil? The lamp makes light ascend from the depths of the earth. The older the substance that does the work, the more surely the lamp is dreamed about in its status as creating creature.

But these reveries on the cosmogonies of the light no longer belong to our age. I recall them here only in order to draw attention to an unknown and lost oneirism, one that has become a matter of history, the knowledge of an old knowledge.

Thus, we wish to guide our dreams according to the inspiration of a great dreamer. If we are attentive to Bosco, we can discover the depths of the reveries of a childhood sustained by its dreams. We enter with Bosco into the labyrinths where memories and dreams intersect. Childhood, from the perspective of its dreams, is unfathomable. We always distort it a

CHAPTER FIVE • 67

little in creating a story. It is sometimes distorted if we dream more, sometimes if we dream less. When he attempts to convey to us the feelings that bind him to the lamp, Bosco is sensitized by these undulations of memories and dreams. A double ontology is then required for us to distinguish what is simultaneously the being of the lamp and the being of him who dreams of remaining faithful to his first illuminations. We touch the roots of poetic feeling for an object charged with memories. Bosco writes: "a feeling that comes to me from that childhood whose solitude, I think, I paraphrase a little ponderously."[4]

III

After such companionship with the lamp as a child, it is not surprising that throughout his work, Bosco makes the lamp a real character with a genuine role in the story of a life. In his many novels, the cozy family lamps mark the humanness of the house and the ongoing life of its family. Often an old servant takes the ancestral lamp in her charge. An old servant prolongs the childhood peace of the master she attended as a child simply by hallowing familiar objects. She knows how to find the right lamp for each great event of domestic life. Such is the old Sidonie who, knowing the hierarchical dignity of a candelabrum, lights all the candles of the silver one for a momentous occasion.

During solemn hours a rustic lamp accentuates through its simplicity the natural drama of life and death. During a somber vigil, when his good servant may have died, the dream hero who is the central character of Bosco's novel *Malicroix* finds moral succor in the lamp.

> For I needed help and, I don't know why, I looked for it in the fire of this little lamp. It gave poor light, being only an ordinary lamp that, badly snuffed, sometimes sparked and threatened to go

---

4. Ibid., 317.

68 • THE FLAME OF A CANDLE

out. Even when its slight flame weakened, it retained a religiously calm clarity. It was a gentle and friendly being that communicated to me in my distress the simple wavy movement of its lamp life. For only a bit of oil fed its glass globe. Unctuous oil that rose in the lamp was dissolved by the flame in its light. But the light, where was it going?[5]

Yes, the light of a gaze, where does it go when death places its cold finger on the eyes of the dying?

IV

Even in life's undramatic hours, lamptime is a solemn time, a time which one must meditate in its slow tempo. One poet, a flame dreamer, knew how to incorporate this slow time in the same sentence that expresses the being of a lamp:

. . . This attentive lamp and the evening take counsel together . . . [6]

The two sets of ellipses are in Fargue's text. Thus the poet enjoins us to speak softly the prelude of an agreement between dim light and the first shadow of evening.

A slow movement unfolds in the chiaroscuro of the dream, a movement that propagates peace: "the lamp extends its soothing hands"[7]; "a lamp extends its wings in the room."[8] It seems that the lamp takes its time to light the whole room progressively. The wings and hands of light move slowly as they brush the walls.

And Léon-Paul Fargue hears the lamp whisper beneath the shell of the lampshade. An ebb and flow of light, both very gentle, lift and settle the luminous covering: "The lamp sings its nimble song as softly as the sound one hears in sea shells."

Octavio Paz also hears the lamp that murmurs:

  5. Bosco, *Malicroix*, 232.
  6. Léon-Paul Fargue, *Poèmes* suivi de *Pour la musique* (Paris: Gallimard, 1963), 114.
  7. Ibid., 134.
  8. Ibid., 111.

CHAPTER FIVE • 69

> . . . faint light from an oil lamp. The light discusses, moralizes, debates with itself. It tells me that no one will ever come.[9]

It seems that silence grows when the lamp speaks low:

> A salt silence made the lamps chime,

says the Belgian poet Roger Brucher.[10]

Images of time that passes by flowing and of time that passes by burning are here brought into harmony. Fargue's lamp is a great image of quiet and slow time. Igneous time, in the lamp's flame, restrains its convulsions. To speak of the lamp's fire one must breathe in peace.

So many of Georges Rodenbach's lamps would impose the same tranquility upon us! In a single line from *Mirror of the Native Sky*,[11] we receive this important lesson:

> Friendly lamp with the slow gaze of a calm fire.

When the evening is come and the lamp is lighted, a poet of lamps lives more than a mechanical moment:

> The room is surprised
> At this lasting happiness.[12]

The lamp imbues the dreamer's room with a joy of light.

We could easily assemble a large number of images that tell in a single stroke the human value of lamps. When these images are good, they have the privilege of simplicity. It seems that the evocation of a lamp is certain to resonate in the souls of readers who love to remember. A poetic halo surrounds lamplight in the chiaroscuro of dreams that reanimate the past.

9. Paz, ¿Aguila o sol?/ *Eagle or Sun*, 107.
10. Roger Brucher, *Vigiles de la rigueur*, 21.
11. Georges Rodenbach, *Le Miroir du ciel natal*, 19.
12. Ibid., 4.

But instead of scattering my proof of the lamp's psychological value over numerous examples, I prefer to call on one of Henri Bosco's most beautiful stories, in which the lamp is the primary mystery of a psychologically mysterious novel. This novel is entitled *Hyacinthe*. In it we find a young woman, the character that Bosco's readers knew as a child in the two stories *The Garden of Hyacinthe* and *Sassy Donkey*. Continuing from one novel to the next, Bosco's characters are thus the oneiric companions of his creative life. To state all that I have in mind, I would add that in Bosco's work the lamp too is an oneiric companion.

What a great task it would be for a psychologist to ferret out from the jumble of dreams and nightmares the personality of this intimate being, this double being who "resembles us like a brother"! Then we would know the unity of being revealed by our dreams. We would truly be the dreamers of ourselves. We would understand others oneirically when we understood the unity of being in their dreaming existence.

But let us look a little more closely at Bosco's lamp in the story of *Hyacinthe*.

V

From the first page, the lamp is Existence. No more than six lines into the book—during which we are told that the narrator is living on a deserted plateau, in a deserted house, in an empty garden bordered by a wall—a lamp intervenes, someone else's lamp, a distant and unexpected lamp. On a first reading one does not suspect in these words of extreme simplicity the nascent drama of solitudes which is given in these few lines:

> In this wall, pierced by a narrow window, the lamp was suddenly lighted for the first time since the evening of my arrival. I was vexed by it.
>
> I waited on the road. I was hoping that someone would pull the window shades. But no one did. The lamp was still shining when I decided to go back home. Since then, I had seen it lit every evening at the first shadows.

CHAPTER FIVE • 71

> Sometimes, very late at night, I went out onto the road. I wanted to know if it was still burning.
> It was there. It was not extinguished until early morning.

We must stop here, for as lamp dreamers we have a problem: that of *the lamp of the other person*. Phenomenologists of the consciousness of the other person have not dealt with such a problem. They do not know that a distant lamp is a sign of the other person.

For a dreamer of the lamp, there are two kinds of lamps belonging to the other person. There is the morning lamp and the evening lamp, the lamp of the First to Rise and the lamp of the Last to Bed. Bosco has increased the problem twofold by confronting the lamp that shines all night. What is this lamp that belongs to the other person; what is this other person with his strange lamp? The entire novel *Hyacinthe* is an answer to these questions.

But to instruct ourselves in the phenomenology of solitude, we must stay with first impressions. Bosco's first page is of extreme sensitivity. The being that came to the empty plateau to seek solitude is troubled by a lamp which burns five hundred meters from his dwelling. The lamp of the other disturbs the repose he finds beside his own lamp. Thus there is a rivalry of solitudes. One wants to be alone to be alone, alone in having a lamp to signify solitude. If the opposing solitary lamp illuminated domestic chores, if it were no more than an implement, then Bosco as meditating dreamer of the lamp is not challenged in any way, nor does he suffer. But two philosopher's lamps in the same village are too many, one too many.

The *cogito* of a dreamer creates its own cosmos, a particular cosmos, a cosmos that is altogether his. His reverie is disturbed, his cosmos marred, if the dreamer knows that another's reverie sets up a world in opposition to his own.

Thus a psychology of inner hostility soon develops in the first pages of *Hyacinthe*. This distant lamp is certainly not "turned in" on itself. It is a lamp that *waits*. It watches so unremittingly that it *guards*. It guards and hence it is malevolent. An entire structure of hostilities is born in the soul of a dreamer whose solitude has just been violated. Thus Bosco's novel turns on a new axis: since the distant lamp guards the plateau, the dreamer troubled by this surveillance will watch the watcher. The dreamer of the lamp hides his own lamp to spy upon the lamp of the other.

I have taken advantage of Bosco's text to present a somewhat studied nuance in the psychology of the lamp. I forced the issue a bit to show how the lamp of the other could awaken our indiscretion, disturb our solitude, and challenge our pride in keeping watch. Every one of these nuances, though a little forced, awakens the idea that the lamp, like all values, can be affected by ambivalence.

But in the novel that begins with a frustration of solitude, the stranger's lamp soon becomes helpful—good lamp that it is—to the dreamer who tells Bosco's story. The dreamer then dreams of the solitude of the other who is to be comforted. The sudden turn occurs as early as page 17:

> It is then that the distant lamp suddenly took on unexpected importance. Not that its brightness could have become more vivid in the heart of this precocious darkness,[13] because it always shone with the same gentleness, but the light it gave seemed more familiar. One could say perhaps that the mind whose work or reverie it illuminated now found in its light a more friendly warmth and loved its calm presence. In my eyes, it had lost its value as sign, what it promised to one who waited, in order to become the lamp of contemplation.

When snow invades the plateau, when winter stops all life, solitude becomes isolation. The dreamer feels distress. Will he flee "the savage wind-swept plain"? It is in dreaming of the distant lamp that he finds help.

---

13. This scene was written at dusk one winter.

CHAPTER FIVE • 73

On the snow-covered plain,

> I saw the lamp; it held me. I watched it now with a silent tenderness. Someone had lighted it for me; it was my lamp. I began to imagine as my equal the man that watched over it so late in the night, in his warm light. Sometimes, carried beyond this resemblance, it was myself that I imagined, attentive to some meditation which, however, was yet impenetrable for me.[14]

The dreamer's trusting reaction to the distant lamp has not reached its bounds. The word "impenetrable" indicates a repressed question. The wavering of trust and mystery has not ceased. In order to find repose, beyond all psychological mysteries, one must become the person who keeps his vigil by the lamp. All meditation tends toward this desire: "Behind the lamp resides this soul—this soul that I would have wanted to be."

I have given only a small measure of the rich variations in Bosco's work which give life to the reverie of the other person's lamp. But then, even if I were to comment on Bosco's thirty pages line by line, would I be able to point out objectively a beauty that is by turns delicate and profound? I have read and reread *Hyacinthe* often. Never have I read it the same way twice . . . What a bad professor of literature I would have made! I dream too much when reading. I also remember too much. With each reading I encounter incidents of personal reverie, incidents of memory. A word, a gesture, stops my reading. Bosco's narrator pulls his window shades in order to hide the light; I remember evenings when I did the same thing, in a house of long ago. The carpenter of the village had cut two hearts in the middle of the shutters so that the morning sun could still awaken the household. Thus in the evening and late at night, through the two openings in the shades, the lamp, my lamp, threw two hearts of golden light over the sleeping countryside.

14. Bosco, *Hyacinthe*, 1946, 18.

Epilogue

My Lamp and My Blank Page

I

BY RECALLING A distant past spent in work, by re-imagining the numerous but monotonous images of an obstinate worker reading and meditating under the lamp, one undertakes to live as though he were the sole character in a painting: a room whose walls are hazy and converge toward the center, concentrated around the meditator seated at a table lit by a lamp. During a long life, the picture has admitted a million variations. But it retains its unity, its central life. It is now a stable image on which memories and dreams are founded. The being who dreams concentrates on it so that he can remember the being who used to work. Is it comfort, is it nostalgia to recall the little rooms in which one used to work, where one had the energy to work well? The real space of solitary work is, in a small room, the circle lit by the lamp. Knowing as much, Jean de Bosschère wrote: "Only a narrow room permits work."[1] And the work light makes the entire room take on the dimensions of the table. It is thus that, in my memories, the lamp of long ago focuses my dwelling place and refashions the solitude of my will, my worker's solitude!

The worker under the lamp is thus a *primordial engraving*, valid for me because of a thousand memories and, at least as I imagine it, valid for all. This picture, I am sure, needs no caption. We do not know what this lamplight worker is thinking, but we do know that he is thinking, that he is alone and

---

1. Jean de Bosschère, *Satan l'Obscur*, 195.

thinking. This primordial engraving bears the mark of solitude, the characteristic mark of a type of solitude.

How much better, how well I would work, if I could find myself again in one or another of my "primordial" engravings!

II

Solitude increases if, on the table lit by the lamp, the solitude of the blank page is displayed. The blank page! A great desert to cross, never before crossed. This blank page that remains blank during each vigil, is it not the great sign of a solitude that endlessly begins again? And what a solitude dogs the solitary man when it is the solitude of a worker who not only wants to learn, who not only wants to think, but who *wants to write*! Then the blank page is a nothingness, a painful nothingness, the nothingness of writing.

Yes, if only one could write! After that, perhaps one could think. *Primum scribere deinde philosophari*, according to a witticism of Nietzsche.[2] But one is much too alone to write. The blank page is too blank, too empty at first for one to begin truly existing by writing. The blank page imposes silence. It contradicts the familiarity of the lamp. From this point on the "engraving" has two polarities: the polarity of the lamp and of the blank page. The solitary worker is torn between these two. A hostile silence thus reigns in my "engraving." Did Mallarmé not live in a divided "engraving" when he recalled

> . . . The deserted light of the lamp
> On the empty paper which its whiteness protects.[3]

III

How good it would be—and generous to oneself—to start everything over again, to begin living by writing! To be born

---

2. Nietzsche, *The Gay Science*, trans. W. Kaufman (New York, 1974), 54: "First write, then philosophize."
3. Mallarmé, "Sea Breeze" in *Poems*, trans. Roger Fry (New York, 1951), 55.

in writing, to be born by writing: the great ideal of the great solitary vigils! But in order to write in the solitude of one's being, as though one had had the revelation of a blank page of life, *adventures in consciousness* are required, adventures in solitude. But can consciousness by itself give variety to its solitude?

Yes, how does one experience adventures in consciousness by remaining alone? Could one discover adventures in consciousness by descending into his own depths? How many times, living in one of my "engravings," I believed that I was deepening my solitude. I believed that I was descending, spiral after spiral, the stairway of being. But now I see that in such descents, though I believed I was thinking, I was really dreaming. Being is not below. It is above, always above—precisely when at work in solitary thought. Thus, in order to be reborn into the full youth of awareness as one sits before the blank page, one must add a little more shadow to the chiaroscuro of old, faded images. It would be necessary in turn to re-engrave the engraver—to re-engrave in each vigil the very existence of the solitary man in the solitude of his lamp—in short, to see everything, think everything, say everything, write everything from the perspective of primordial existence.

IV

In short, in every account of life's experiences—experiences that are torn apart and that tear apart—it is rather mostly before my blank paper, before the blank page placed on the table at the right distance from my lamp, that I am truly at my *table of existence.*

Yes, it is at my table of existence that I have known maximum existence, existence in tension—tending toward an "ahead," toward a "further ahead," toward an "above." All around me is rest, tranquility; only my Being-seeking being strains in its improbable need to be another being, a more-than-being. And thus it is that with Nothing, with Reveries, one believes one will be able to make books.

But when a small album of the chiaroscuros of a dreamer's psyche is finished, the time of nostalgia for strictly ordered thoughts returns. By pursuing my candle romanticism I have presented only a part of life before the table of existence. After so many reveries I am seized with haste to instruct myself further, and consequently to put aside the blank paper in order to study a book, a difficult book, one always a little too difficult for me. Through this tension in facing a rigorously argued book, the mind is constructed and reconstructed. Every evolution of thought, every prospect for thought, occurs in a reconstruction of the mind.

But is it time again for me to find the worker whom I know so well and return him to my engraving?

Endnotes

Like many of his contemporaries in France, Gaston Bachelard documented his sources in a rather desultory fashion, footnoting or not at his discretion. This practice, so disconcerting to American readers accustomed to stricter conventions, was exacerbated by wartime conditions. Yet the very casualness of Bachelard's documentation may well contribute to the characteristic flow of his prose. For this reason, the author's original notes, corrected when possible, have been kept in place at the bottom of the page. All other information or commentary has been assembled in these endnotes. They are referenced neither by means of superscripts in the text nor by page and line numbers but by page and keyword or phrase.

The reader can locate a keyword in the text by scanning the left margin. For instance, a note on Bachelard's use of the word *pancalism* on page three is listed after the key phrase: "world's beauty . . . pancalizing."

P.3   **world's beauty . . . pancalizing:** *Pancalism* is explained in a footnote to *Air and Dreams* (Dallas, 1988), 49. "To answer the objections that some have made over my use of the word 'pancalism,' let me point out that I borrowed it from Baldwin's terminology. I have used it to express the idea that pancalistic activity tends to transform any contemplation of the universe into an affirmation of universal beauty." Cf. James Mark Baldwin, "Pancalism: A Theory of Reality," in *Genetic Theory of Reality* (New York and London, 1915), ch. XV, 275 ff.

P.3   **title *A Poetics of Fire*:** Bachelard never completed *The Poetics of Fire*. In 1988 Suzanne Bachelard edited and published *Fragments d'une poétique du feu*. The Dallas

Institute Publications will publish Kenneth Haltman's English translation in 1989.

P.6 **[which blaze up and go out, you know not why, in different parts]:** This phrase was not included in the French text.

P.6 **artists— . . . "describe":** In the French the common root of *describe, write, inscribe* and *transcribe* is apparent: ". . . ce clair-obscur, comment, non pas le peindre—c'est là le privilège des grands artistes—mais le 'décrire'? Comment l'écrire? Nous voulons nous-même aller plus loin: ce clair-obscur, comment l'*inscrire* dans le psychism . . ."

P.13 **Martin Kaubish:** full citation reads "Martin Kaubish, *Anthologie de la poésie allemande*, trad. René Lasne et Georg Rabuse, t.II."

P.24 **trous.":** "bougies à trous," cratered candles.

P.32 **edition of the *Metamorphoses of the Soul*:** The English title is *Symbols of Transformation* (see footnote 11 in this chapter).

P.39 **Octavio Paz, *Eagle or Sun?*:** Trans. Eliot Weinberger (New York, 1976), 127.

P.44n **13. Cf. . . . excrement of a flame:** Bachelard is referring to Novalis, *Schriften*, Band 3 (Leipzig: Bibliographisches Instituten, n.d.), 40.

P.53 **Poplar of fire, water jet:** Paz' phrase, "Chopo de fuego, chorro de agua," [¿*Aguila o Sol?* (Mexico, 1951), 114] is from the poem "Himno Futuro" and is not included in ¿*Aguila o Sol?/Eagle or Sun*, trans. Eliot Weinberger (New York: 1976); a note states that "both the Spanish and English texts have undergone major revisions" since 1968.

P.61 **through . . . Champagne:** In French the phrase "d'un oeil champenois" is idiomatic for "with a sense of humor." Since Bachelard himself was a native of

P.63   Champagne, we chose to retain the literal meaning in our translation.

P.63   **Tagore:** Rabindranath Tagore, *Fireflies* (New York, 1928), 29. A note at the head of the volume states "*Fireflies* had their origin in China and Japan where thoughts were very often claimed from me in my handwriting on fans and pieces of silk" (5). Bachelard's citation is incomplete. The whole poem reads: "Bigotry tries to keep truth safe in its hand/with a grip that kills it./Wishing to hearten a timid lamp/great night lights all her stars."

The contrast between lamp and the heavens is frequent in Tagore's writings.

P.68   **rose in the lamp:** The 1948 edition of *Malicroix* has "qui montait à la flamme" rather than "qui montait à la lampe."

Author/Title Index

A

Air and Dreams 9, 40, 79
Anthologie de la poésie allemand 13, 40n
Anthologie 49
Arnold, Céline 49
Asselineau, Roger 41n

B

Baader, Franz von 28, 29
Bachelard, Gaston 8n
Balzac, Honoré de 50, 51
Banville, Théodore de 26
Béguin, Albert 17n, 23
Bosco, Henri 11, 65-68, 70-73
Bosschère, Jean de 24, 25, 63, 75
Bourdeillette, Jean 56
Brucher, Robert 69

C

Camoens, Luiz 26, 27
Cassou, Jean 36
Caubère, Jean 52
Claudel, Paul 41, 44
Consuelo 5, 6
Conversations de Goethe et d'Eckermann 42n
Corbin, Henry 45, 59, 60

D

D'Annunzio, Gabriele 55, 58, 59
Desoille, Robert 9
Dickens, Charles 54
Dictionary of French Onomatopoeia 28
Divan, The 33, 34

E

Eagle or Sun? 39, 79
Eckermann 42
Edsman, Carl-Martin 42

Emié, Louis 23
Empedocles 34
Eye Listens, The 41

F

Faraday, Michael 47
Fargue, Léon-Paul 68, 69
Fire 58
Fireflies 63
"Flame" 41
Franz von Baader et la connaissance mystique 28n
Frazer, James 61

G

Garden of Hyacinthe, The 70
Garnier, Pierre 47n
Goethe, Johann Wolfgang von 33, 34, 42, 44
Gubernatis, Angelo de 27n
Guillaume, Louis 51, 52

H-I-J

History of a Candle 47
Hugo, Victor 50, 51
Hyacinthe 70, 71, 73
Ignis divinus 42
Inferno 30, 31
Isis 41
Jabès, Edmond 40n
Joubert, Joseph 15, 16
Jouve, Pierre-Jean 34, 35
Jung, Carl 17, 32, 33

K-L

Kaubish, Martin 13, 79
L'Isle-Adam, Villiers de 41
Lavoisier, Antoine 8
Le nom du feu 23
Le Sieur de La Chambre 27n
Lichtenberg, George 23

Lichtenberger, Henri 34
"Lighted Candle, The" 23
Lundkvist, 58

M

Maeterlinck 44
Malicroix 67, 68, 80
Mallarmé 76
Mandiargues, Pieyre de 57
Matisse, Henri 54
Memoirs of a Tourist 24
Metamorphoses of the Soul 32, 79
Michelet, V.-E. 21
Milosz, Oscar Vladislas 36
Minkowski, Eugene 64
Mirror of the Native Sky 69

N

Nietzsche, Friedrich 45n
Nodier, Charles 28, 29, 46, 47
Novalis 9, 10, 42-45, 47, 49, 55, 66, 79
Novices of Sais, The 43n

O-P

"Old Oak, The" 51
Paracelsus 17
Paulina 1880 34, 35
Paz, Octavio 39, 53, 68, 69, 79, 80
Pensées 15, 15n
Plato 34
Plotinus 34
Poetics of Fire, A 3
Poetics of Space, The 11

Poetry of Flames, The 3
Poetry of Light, The 54
Proclus 59, 60

R-S

Rachilde 54
Rembrandt 6
Rodenbach, Georges 58, 69
Saint-Martin, Claude de 43
Saint-Pierre, Bernardin de 56, 57
Sand, George 5, 6, 16
Sassy Donkey 70
"Song of the Moth, The" 32
Soreil, Arsène 54
Stendhal 24
Strindberg, August 30, 31
Susini, Eugene 28
"Sympathy and Theopathy among the 'Fidèles d'amour' in Islam" 59

T

Tagore, Rabindranath 63, 80
Thiry, Marcel 54
Toward a Cosmology 64
Trakl 40
Treatise on Fire and Salt 18
Tzara, Tristan 25, 25n

V-Z

Vigenère, Blaise de 18-20, 49
Wahl, Jean 5
West-eastern Divan 44n
Zohar 18, 20

Subject Index

A

anima 7, 16
animality 44
animus 7, 16
Aristotelian philosophy 21
ascension 54
ascensional 9, 10, 39, 40, 43; ascensional being 43; ascensional force 39, 40

B

Bergsonian instant 64
bird 47
Bonifazio 59
butterfly 33-35

C

Caubère's water jet 52
Chardin 56
chiaroscuro 4-7, 36, 55, 68, 69, 77; of dream 36, 69
childhood 11, 65-67
clignoter 28, 29
coal 57, 58
companionship 23, 63, 65, 67
consciousness 5, 6, 19, 25, 32, 33, 36, 71, 77; chiaroscuro of 5
contemplation 20
cornflower 58
correspondence 51; Baudelairean 51, 59
creature 32, 45, 65, 66

D

dahlia 56
death 13, 16, 17, 31-35, 67, 68
destiny 3, 14, 19, 21, 32, 33, 35, 46, 52
double ontology 67
dream 65; of flight 9; philosophers' 10; waking 9
dynamism 21, 25, 41

E

elan 40, 45
Empedocles 34, 35
Empedocles complex 34
Eros 34, 35

F

fantasy 16, 18
fire birds 46
fire 3, 7, 8, 13, 15, 18, 19, 22, 24, 25, 27, 30, 31, 34, 35, 39-46, 49-61, 65, 67, 69; and water 52; as light 41; physics of 8, 13; unity of 56
fireplace 46, 57, 58
flame thinker 19, 29
flame dreamer 29, 68
flame *and*: animal life 43; flower 49, 53, 55, 56; torch 21; water jet 52
flame: animalized 47; flowering 44; as imagination 10; liquid 30; white 18, 19, 21, 49; yellow 19, 21
flame-flowers 56
flame-life 45
flower-flames 55, 56

G

Gehenna 28, 29
geraniums 57
German romanticism 42
grape 54

H

heliopathy 60
heliotrope 49, 59, 60
Heracles 52
hydrangea 59
hymn 53
hypervalorization 19

I

ignis divinus 42
image 20; ascensional 9; dream 7; literary 3, 4; metaphor 1; poet's 58; and poetic reverie 2; spoken 4
image-seed 51, 52
image-thought 15
imagination 1; and memory 8; literary 3, 10
implementarity 65

K-L

Koran 60
lamp 10, 11, 18, 29-31, 34, 37, 43, 51, 54, 56-58, 63-73, 75-77; and solitude 71; of the other 71, 72
lamp-bearers 54
Laocoon 52
lightbulb 63, 64
liturgy 21, 41

M

matter 3, 18, 20, 29, 31, 41, 44, 45, 66
meditation 1, 3, 11, 14, 16, 17, 35, 45, 60, 73
melancholy 26, 53
memory 2, 8, 11, 23, 26, 36, 46, 65-67, 73, 75; and dream 66
Menri 61
metaphor 20
Miss Miller 32
model-phenomenon 20
"moist fire" 15
moth 32, 34

N-O

nightmare 4, 6, 7, 23, 70
nostalgia 33
oil 49, 63, 64, 66, 68, 69
oneiric personality 24
oneiric objects 23
oneiric companion 70
oneiric 14, 23, 24, 70
oneirism 10, 44, 66
ontophany 59

P

pancalizing 3
Paulina 34, 35
penumbra 6, 55, 65
Persia 56
phenomenologist 64, 65, 71
phenomenology 42, 71; of solitude 71
phoenix 47
phototropism 34
physics 42
plant life 49, 50
plant flames 50, 57
Plato 34
Plotinus 34
poet-dreamer 55
poetics 3, 11, 43, 45, 55; of material elements 43
primal speech 53
primary image 36, 46
Promethean 24
psyche 10; aesthetic values of 5; chiaroscuro of 4-6, 36
"Psychic height" 3
psychology: intersubjective 7; of family 11; of the familiar 7; of the house 11; of the unconscious 4

R-S

reverie 6, 15, 16; of familiar 7; law of 2; and melancholy 26; of memories 65; and noctural dream 7; of the other 11, 73; poetic 2; poetic aspect of 6; primal 2, 14; verticalizing 39
rose 57, 58, 68
Satan 25, 75
schizophrenia 31
schrack 28, 29
seed-image 51
Sidonie 67
solitude 9, 11, 14, 23, 25-27, 31, 35-37, 56, 67, 71, 72, 75, 76, 77
Sufi 34
sun 17, 32, 34, 39, 54-56, 59, 69, 73
surrealism 3, 52
symbol 20, 21, 22, 24, 25, 32, 34, 41

T

table 13, 16, 20, 27, 37, 40, 56, 75-78
Thanatos 35
thought-image 15
thoughts 1, 5, 14, 15, 18, 20, 25, 37, 51
torch 18, 21, 34, 41, 57, 58
tree 44, 46, 49-56
tulip 56

U-V-W

unconscious 4, 5
unity of being 70
valorization 19
value 4, 5, 15, 19-22, 31, 39, 40, 42-43, 51, 53, 57, 61, 64, 66, 69, 70, 72; antivalue 19; cosmology of 42; poetic 61
vertical 3, 14, 15, 24, 39, 41, 43, 49, 53, 58
verticality 9, 10, 19, 39-41, 43; and destiny 19
verticalizing 3, 39, 40; will 40
vesania 31
vigil 14, 16, 26-28, 37, 67, 73, 76, 77
water jet 52, 53
wick 17, 41, 47, 57, 64, 66